4th BioASQ Workshop 2016

A Challenge on Large-Scale Biomedical Semantic Indexing and Question Answering

Held at ACL 2016

Berlin, Germany
13 August 2016

ISBN: 978-1-5108-2764-6

Printed from e-media with permission by:

Curran Associates, Inc.
57 Morehouse Lane
Red Hook, NY 12571

CURRAN ASSOCIATES INC. proceedings.com

Some format issues inherent in the e-media version may also appear in this print version.

Printed by Curran Associates, Inc. (2016)

Association for Computational Linguistics
209 N. Eighth Street
Stroudsburg, Pennsylvania 18360

Phone: 1-570-476-8006
Fax: 1-570-476-0860

acl@aclweb.org

Additional copies of this publication are available from:

Curran Associates, Inc.
57 Morehouse Lane
Red Hook, NY 12571 USA
Phone: 845-758-0400
Fax: 845-758-2633
Email: curran@proceedings.com
Web: www.proceedings.com

ACL 2016

The 4th BioASQ Workshop
A challenge on large-scale biomedical semantic indexing and question answering

Proceedings of the Workshop

August 13, 2016
Berlin, Germany

Production and Manufacturing by
Taberg Media Group AB
Box 94, 562 02 Taberg
Sweden

Preface

The fourth BioASQ workshop on biomedical semantic indexing and question answering took place in Berlin, Germany on August 13th, 2016 and was hosted by the Humboldt University. The workshop was supported by the BioASQ project[1], which organizes the corresponding annual challenge. The goals of the workshop were to present the results of the fourth BioASQ challenge and further the interaction with the wider community of biomedical semantic indexing and question answering. The presenters represented research teams from different parts of the globe and with different viewpoints to the problem. This contributed to a very lively and interesting discussion among the participants of the workshop. Six papers were presented during the workshop. All were selected by peer review for presentation. This volume includes 7 papers: The first paper gives an overview of the challenge, including especially the datasets that were used throughout the challenges and the overall results achieved by the participants. The remaining six papers are those presented at the workshop. The first of these papers is a new extension of the MTI system. In particular, Learning to Rank methodology is used as a boosting component of the MTI system. The second paper present a system which includes several multi-label classifiers (MLC) that are combined in ensembles. Elastic-Search for indexing is the object of discourse of the third workshop paper. The fourth paper presents a system which uses TmTool in addition to MetaMap, to identify possible biomedical named entities, especially out-of-vocabulary concepts. They also introduced a unified classification interface for judging the relevance of each retrieved concept, document, and snippet, which can combine the relevant scores evidenced by various sources. The system presented in the fifth paper relies on the Hana Database for text processing. It uses the Stanford CoreNLP package for tokenizing the questions. Each of the tokens is then sent to the BioPortal and to the Hana database for concept retrieval. The concepts retrieved from the two stores are finally merged to a single list that is used to retrieve relevant text passages from the documents at hand. The last paper focuses on the retrieval of relevant documents and snippets. The proposed system uses a cluster-based language model. Then, it reranks the retrieved top-n sentences using five independent similarity models based on shallow semantic analysis.

We wish to thank all who participated to the success of this workshop, especially the authors, reviewers, speakers and participants.

Ioannis A. Kakadiaris, George Paliouras and Anastasia Krithara
August 2016

[1] http://www.bioasq.org

Organizers:

Ioannis A. Kakadiaris, University of Houston, USA
George Paliouras, NCSR "Demokritos", Greece and University of Houston, USA
Anastasia Krithara, NCSR "Demokritos", Greece

Program Committee:

Ion Androutsopoulos, Athens University of Economics and Business, Greece
Nicolas Baskiotis, Université Pierre et Marie Curie, France
Dimitris Galanis, National Technical University of Athens, Greece
Brigitte Grau, LIMSI/CNRS, France
Aris Kosmopoulos, NCSR "Demokritos", Greece
Zhiyong Lu, National Library of Medicine, USA
Prodromos Malakasiotis, Athens University of Economics and Business, Greece
Jim Mork, National Library of Medicine, USA
Diego Molla, Macquarie University, Australia
Henning Müller, University of Applied Sciences, Switzerland
Claire Nedellec, INRA, France
Mariana Neves, University of Potsdam, Germany
Harris Papageorgiou, ILSP, Greece
Ioannis Partalas, Viseo group, France
John Prager, Thomas J. Watson Research Center, IBM, USA
Francisco J. Ribadas-Pena, University of Vigo, Spain
Hagit Shatkay, University of Delaware, USA
Grigoris Tsoumakas, Aristotle University of Thessaloniki, Greece
Christina Unger, Bielefeld University, Germany
Ellen Voorhees, National Institute of Standards and Technology, USA

Invited Speaker:

Sherri Matis-Mitchell, independent consultant for Text, Data and Social Media Analytics at Data Star Insights

Table of Contents

Conference Program

Results of the 4th edition of BioASQ Challenge
Anastasia Krithara, Anastasios Nentidis, Georgios Paliouras and Ioannis Kakadiaris

Using Learning-To-Rank to Enhance NLM Medical Text Indexer Results
Ilya Zavorin, James Mork and Dina Demner-Fushman

LABDA at the 2016 BioASQ challenge task 4a: Semantic Indexing by using ElasticSearch
Isabel Segura-Bedmar, Adrián Carruana and Paloma Martínez

Learning to Answer Biomedical Questions: OAQA at BioASQ 4B
Zi Yang, Yue Zhou and Eric Nyberg

HPI Question Answering System in BioASQ 2016
Frederik Schulze, Ricarda Schüler, Tim Draeger, Daniel Dummer, Alexander Ernst, Pedro Flemming, Cindy Perscheid and Mariana Neves

KSAnswer: Question-answering System of Kangwon National University and Sogang University in the 2016 BioASQ Challenge
Hyeon-gu Lee, Minkyoung Kim, Harksoo Kim, Juae Kim, Sunjae Kwon, Jungyun Seo, Yi-Reun Kim and Jung-Kyu Choi

Large-Scale Semantic Indexing and Question Answering in Biomedicine
Eirini Papagiannopoulou, Yiannis Papanikolaou, Dimitris Dimitriadis, Sakis Lagopoulos, Grigorios Tsoumakas, Manos Laliotis, Nikos Markantonatos and Ioannis Vlahavas

Results of the 4th edition of BioASQ Challenge

Anastasia Krithara[1], **Anastasios Nentidis**[1], **George Paliouras**[1], and **Ioannis Kakadiaris**[2]

[1]National Center for Scientific Research "Demokritos", Athens, Greece
[2]University of Houston, Texas, USA

Abstract

The goal of this task is to push the research frontier towards hybrid information systems. We aim to promote systems and approaches that are able to deal with the whole diversity of the Web, especially for, but not restricted to, the context of biomedicine. This goal is pursued by the organization of challenges. The fourth challenge, as the previous challenges, consisted of two tasks: semantic indexing and question answering. 16 systems participated by 7 different participating teams for the semantic indexing task. The question answering task was tackled by 37 different systems, developed by 11 different teams. 25 of the systems participated in the phase A of the task, while 12 participated in phase B. 3 of the teams participated in both phases of the question answering task. Overall, as in previous years, the best systems were able to outperform the strong baselines. This suggests that advances over the state of the art were achieved through the BIOASQ challenge but also that the benchmark in itself is very challenging. In this paper, we present the data used during the challenge as well as the technologies which were at the core of the participants' frameworks.

1 Introduction

The aim of this paper is twofold. First, we aim to give an overview of the data issued during the BioASQ challenge in 2016. In addition, we aim to present the systems that participated in the challenge and for which we received system descriptions, as well as evaluate their performance. To achieve these goals, we begin by giving a brief overview of the tasks, including the timing of the different tasks and the challenge data. Thereafter, we give an overview of the systems which participated in the challenge and provided us with an overview of the technologies they relied upon. Detailed descriptions of some of the systems are given in lab proceedings. The evaluation of the systems, which was carried out by using state-of-the-art measures or manual assessment, is the last focal point of this paper. The conclusion sums up the results of this challenge.

2 Overview of the Tasks

The challenge comprised two tasks: (1) a large-scale semantic indexing task (Task 4a) and (2) a question answering task (Task 4b).

Large-scale semantic indexing. In Task 4a the goal is to classify documents from the PubMed[1] digital library into concepts of the MeSH[2] hierarchy. Here, new PubMed articles that are not yet annotated are collected on a weekly basis. These articles are used as test sets for the evaluation of the participating systems. As soon as the annotations are available from the PubMed curators, the performance of each system is calculated by using standard information retrieval measures as well as hierarchical ones. The winners of each batch were decided based on their performance in the Micro F-measure (MiF) from the family of flat measures (Tsoumakas et al., 2010), and the Lowest Common Ancestor F-measure (LCA-F) from the family of hierarchical measures (Kosmopoulos et al., 2013). For completeness several other flat and hierarchical measures were reported (Balikas et al., 2013). In order to provide an on-line and large-scale scenario, the task was divided into three independent batches. In each batch 5 test

[1]http://www.ncbi.nlm.nih.gov/pubmed/
[2]http://www.ncbi.nlm.nih.gov/mesh/

Proceedings of the 54th Annual Meeting of the Association for Computational Linguistics, pages 1–7,
Berlin, Germany, August 7-12, 2016.

sets of biomedical articles were released consecutively. Each of these test sets were released in a weekly basis and the participants had 21 hours to provide their answers. Figure 1 gives an overview of the time plan of Task 4a.

Biomedical semantic QA. The goal of task 4b was to provide a large-scale question answering challenge where the systems should be able to cope with all the stages of a question answering task, including the retrieval of relevant concepts and articles, as well as the provision of natural-language answers. Task 4b comprised two phases: In phase A, BIOASQ released questions in English from benchmark datasets created by a group of biomedical experts. There were four types of questions: "yes/no" questions, "factoid" questions,"list" questions and "summary" questions (Balikas et al., 2013). Participants had to respond with relevant concepts (from specific terminologies and ontologies), relevant articles (PubMed articles), relevant snippets extracted from the relevant articles and relevant RDF triples (from specific ontologies). In phase B, the released questions contained the correct answers for the required elements (articles and snippets) of the first phase. The participants had to answer with *exact* answers as well as with paragraph-sized summaries in natural language (dubbed *ideal* answers).

The task was split into five independent batches. The two phases for each batch were run with a time gap of 24 hours. For each phase, the participants had 24 hours to submit their answers. We used well-known measures such as mean precision, mean recall, mean F-measure, mean average precision (MAP) and geometric MAP (GMAP) to evaluate the performance of the participants in Phase A. The winners were selected based on MAP. The evaluation in phase B for the ideal answers was carried out manually by biomedical experts on the answers provided by the systems. For the sake of completeness, ROUGE (Lin, 2004) is also reported. For the exact answers, we used accuracy for the yes/no questions, mean reciprocal rank (MRR) for the factoids and mean F-measure for the list questions.

3 Overview of Participants

3.1 Task 4a

In this subsection we describe the proposed systems which have sent a description and stress their key characteristics.

In (Papagiannopoulou et al., 2016) flat classification processes were employed for the semantic indexing task. In particular, they used as a training set the last 1 million articles and kept the last 50 thousand as a validation set. Pre-processing of the articles was carried out by concatenated the abstract and the title. One-grams and bi-grams were used as features, removing stop-words and features with less than five occurrences in the corpus. The tf-idf representation has been used for the features. The proposed system includes several multi-label classifiers (MLC) that are combined in ensembles. In particular, they used the Meta-Labeler, a set of Binary Relevance (BR) models with Linear SVMs and a Labeled LDA variant, Prior LDA. All the above models were combined in an ensemble, using the MULE framework, a statistical significance multi-label ensemble that performs classifier selection.

The approach proposed by (Segura-Bedmar et al., 2016) is based on Elastic Search. They use ElasticSearch in order to index the training set provided by the BioASQ. Then, each document in the test set is translated into a query, that is fired against the index built from the training set, returning the most relevant documents and their MeSH categories. Finally, each MeSH category is ranked using a scoring system based on the frequency of the category and the similarity of relevant documents, which contain the category, with the test document to classify.

Baselines. During the challenge three systems were served as baseline systems. The first baseline is a state-of-the-art method called Medical Text Indexer (MTI) (Mork et al., 2014) which is developed by the National Library of Medicine[3] and serves as a classification system for articles of MEDLINE. MTI is used by curators in order to assist them in the annotation process. The second baseline is an extension of the system MTI with the approaches of the first BioASQ challenge's winner (Tsoumakas et al., 2013). The third one, dubbed BioASQ_Filtering (Zavorin et al., 2016) is

[3]http://ii.nlm.nih.gov/MTI/index.shtml

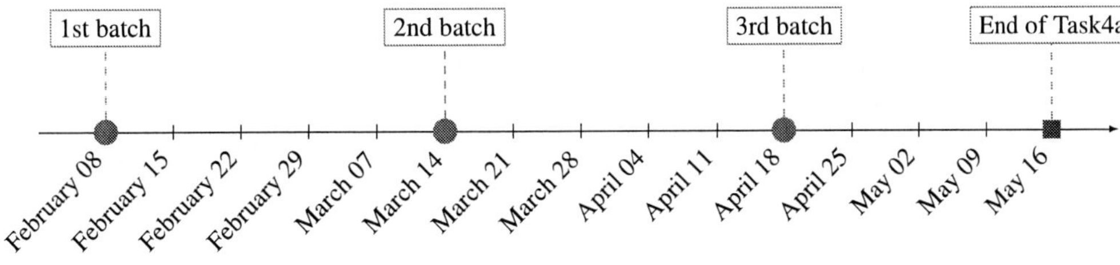

Figure 1: The time plan of Task 4a.

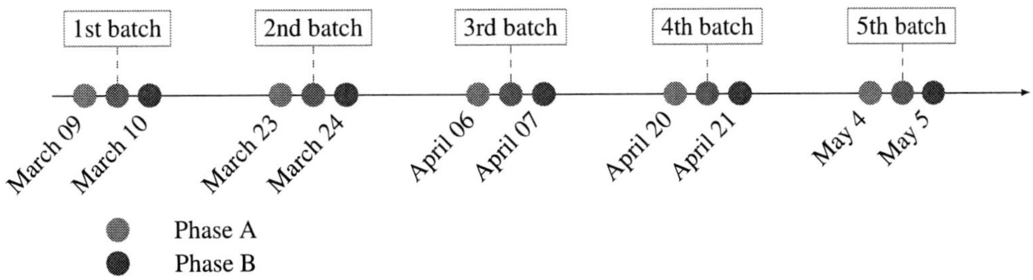

Figure 2: The time plan of Task 4b. The two phases for each batch run in consecutive days.

a new extension of the MTI system. In particular, Learning to Rank methodology is used as a boosting component of the MTI system. The improved system shows significant gains in both precision and recall for some specific classes of MeSH headings.

3.2 Task 4b

As mentioned above, the second task of the challenge is split into two phases. In the first phase, where the goal is to annotate questions with relevant concepts, documents, snippets and RDF triples 9 teams with 25 systems participated. In the second phase, where teams are requested to submit exact and paragraph-sized answers for the questions, 5 teams with 12 different systems participated.

The system presented in (Papagiannopoulou et al., 2016) is based on Indri search engine, and they use MetaMap and LingPipe to detect the biomedical concepts in local ontology files. For the relevant snippets, they calculate the semantic similarity between each one of the sentences and the query (expanded with synonyms) using a semantic similarity measure. Concerning phase B, They provided exact answers only for the factoid questions. Their system is based on their previous participation in BioASQ challenge (Papanikolaou et al., 2014). The system tries to extract the lexical answer type by manipulating the words of the question. Then, the relevant snippets of the question which are provided as inputs for this tasks are processed with the 2013 release of MetaMap in order to extract candidate answers. This year, they have extended their approach by expanding both the scoring mechanism, as well as the set of candidate answers.

The system presented in (Yang et al., 2016), extends the system in (Yang et al., 2015). In particular, they used TmTool (CH et al., 2016), in addition to MetaMap, to identify possible biomedical named entities, especially out-of-vocabulary concepts. In addition, they also extract frequent multi-word terms from relevant snippets to further improve the recall of concept and candidate answer text extraction. They also introduced a unified classification interface for judging the relevance of each retrieved concept, document, and snippet, which can combine the relevant scores evidenced by various sources. A supervised learning method is used to rerank the answer candidates for factoid and list questions based on the relation between each candidate answer and other candidate answers.

The system presented in (Schulze et al., 2016) relies on the Hana Database for text processing. It uses the Stanford CoreNLP package for tokenizing the questions. Each of the tokens is then sent to the BioPortal and to the Hana database for concept retrieval. The concepts retrieved from the two stores are finally merged to a single list that is used to retrieve relevant text passages from the documents at hand. The second system relies on existing NLP functionality in the IMDB. They have extended it with new functions tailored specifically to QA.

The approach presented in (gu Lee et al., 2016) participated in phase A of task 4b. The main focus was the retrieval of relevant documents and snippets. The proposed system uses a clusterbased language model. Then, it reranks the retrieved top-n sentences using five independent similarity models based on shallow semantic analysis.

4 Results

4.1 Task 4a

During the evaluation phase of the Task 4a, the participants submitted their results on a weekly basis to the online evaluation platform of the challenge[4]. The evaluation period was divided into three batches containing 5 test sets each. 7 teams were participated in the task with a total of 16 systems. For measuring the classification performance of the systems several evaluation measures were used both flat and hierarchical ones (Balikas et al., 2013). The micro F-measure (MiF) and the Lowest Common Ancestor F-measure (LCA-F) were used to asses the systems and choose the winners for each batch (Kosmopoulos et al., 2013). 12, 208, 342 articles with 27, 301 labels (19.4GB) were provided as training data to the participants. Table 1 shows the number of articles in each test set of each batch of the challenge.

Table 2 presents the correspondence of the systems for which a description was available and the submitted systems in Task 4a. The systems MTI First Line Index, Default MTI, BioASQ_Filtering were the baseline systems used throughout the challenge. Systems that participated in less than 4 test sets in each batch are not reported in the results[5].

According to (Demsar, 2006) the appropriate way to compare multiple classification systems over multiple datasets is based on their average rank across all the datasets. On each dataset the system with the best performance gets rank 1.0, the second best rank 2.0 and so on. In case that two or more systems tie, they all receive the average rank.

Tables 3 presents the average rank (according to MiF and LCA-F) of each system over all the test sets for the corresponding batches. Note, that the average ranks are calculated for the 4 best results of each system in the batch according to the rules of the challenge[6]. The best ranked system is highlighted with bold typeface.

Table 4: Statistics on the training and test datasets of Task 4b. All the numbers for the documents, snippets, concepts and triples refer to averages.

Batch	Size	# of documents	# of snippets
training	1307	13.00	17.86
1	100	4.56	6.41
2	100	5.25	6.98
3	100	4.79	6.46
4	100	4.90	7.25
5	97	3.93	6.10
total	1804	10.71	14.77

4.2 Task 4b

Phase A. Table 4 presents the statistics of the training and test data provided to the participants. The evaluation included five test batches. For the phase A of Task 4b the systems were allowed to submit responses to any of the corresponding types of annotations, that is documents, concepts, snippets and RDF triples. For each of the categories we rank the systems according to the Mean Average Precision (MAP) measure (Balikas et al., 2013). The final ranking for each batch is calculated as the average of the individual rankings in the different categories. In tables 6 and 7 some indicative results from batch 1 are presented. The detailed results for Task 4b phase A can be found in `http://participants-area.bioasq.org/results/4b/phaseA/`.

Phase B. In the phase B of Task 4b the systems were asked to report exact and ideal answers. The systems were ranked according to the manual evaluation of ideal answers by the BioASQ experts (Balikas et al., 2013), and according to automatic measures for the exact answers.

Table 7 shows the results for the exact answers for the first batch of task 4a. In case that systems

[4] `http://participants-area.bioasq.org/`

[5] According to the rules of BioASQ, each system had to participate in at least 4 test sets of a batch in order to be eligible for the prizes.

[6] `http://participants-area.bioasq.org/general_information/Task4a/`

Table 1: Statistics on the test datasets of Task 4a.

Batch	Articles	Annotated Articles	Labels per article
1	3,740	569	11.25
	2,872	714	12.01
	2,599	275	11.09
	3,294	520	13.72
	3,210	418	11.23
Subtotal	15,715	2,496	11.96
2	3,212	443	10.57
	3,213	371	11.37
	2,831	534	11.78
	3,111	541	10.67
	2,470	268	9.82
Subtotal	14,837	2,157	10.94
3	2,994	89	12.08
	3,044	353	11.79
	3,351	241	10.81
	2,630	93	9.77
	3,130	50	12.56
Subtotal	15,149	826	11.35
Total	45,701	5,479	11.42

Table 2: Correspondence of reference and submitted systems for Task 4a.

Reference	**Systems**
(Papagiannopoulou et al., 2016)	Auth1, Auth2
(Segura-Bedmar et al., 2016)	LABDA ElasticSearch, LargeElasticLABDA, LABDA baseline
Baselines ((Mork et al., 2013),(Zavorin et al., 2016))	MTI First Line Index, Default MTI, BioASQ_Filtering

Table 3: Average ranks for each system across the batches of the task 4a for the measures MiF and LCA-F. A hyphenation symbol (-) is used whenever the system participated in less than 4 times in the batch.

System	Batch 1		Batch 2		Batch 3	
	MiF	LCA-F	MiF	LCA-F	MiF	LCA-F
iria-1	-	-	9.0	9.0	-	-
LABDA ElasticSearch	-	-	-	-	-	-
d33p	-	-	-	-	-	-
auth1	2.75	3.25	3.75	3.75	-	-
Default MTI	4.0	3.0	5.0	4.5	-	-
auth2	-	-	6.0	6.25	-	-
MeSHLabeler	**1.25**	**1.25**	**1.25**	**1.25**	-	-
LargeElasticLABDA	-	-	-	-	-	-
LABDA baseline	-	-	-	-	-	-
BioASQ Filtering	4.5	4.75	5.75	5.5	-	-
MeSHLabeler-2	-	-	2.0	2.0	-	-
MeSHLabeler-1	1.75	1.75	-	-	-	-
MeSHLabeler-3	-	-	3.5	3.25	-	-
CSX-1	-	-	-	-	-	-
MTI First Line Index	5.5	5.75	5.75	6.25	-	-
UCSDLogReg	-	-	-	-	-	-

didn't provide exact answers for a particular kind of questions we used the symbol "-". The results of the other batches are available at `http://participants-area.bioasq.org/results/4b/phaseB/`. From those results we can see that the systems are achieving a very high (> 90% accuracy) performance in the yes/no questions. The performance in factoid and list questions is not as good indicating that there is room for improvements.

5 Conclusion

In this paper, an overview of the fourth BioASQ challenge is presented. As the previous challenges, the challenge consisted of two tasks: semantic indexing and question answering. Overall, as in previous years, the best systems were able to outperform the strong baselines provided by the organizers. This suggests that advances over the state of the art were achieved through the BIOASQ challenge but also that the benchmark in

Table 5: Results for batch 1 for documents in phase A of Task 4b.

System	Mean Precision	Mean Recall	Mean F-measure	MAP	GMAP
testtext	0.169	0.5331	0.2276	0.0981	0.0128
ustb_prir2	0.158	0.5277	0.2164	0.0973	0.0119
ustb_prir4	0.165	0.5254	0.2224	0.0967	0.0109
fdu2	0.147	0.5011	0.2012	0.0885	0.0087
ustb_prir3	0.156	0.497	0.2114	0.0869	0.0095
fdu	0.153	0.5086	0.2081	0.0866	0.0095
ustb_prir1	0.155	0.4936	0.2097	0.0865	0.0088
fdu4	0.15	0.5057	0.205	0.0859	0.012
fdu3	0.154	0.5184	0.2112	0.0849	0.0109
fdu5	0.149	0.4971	0.2036	0.0823	0.01
KNU-SG Team_Korea	0.084	0.2258	0.1065	0.0486	0.0008
HPI-S1	0.1209	0.3266	0.1547	0.0474	0.0012
Auth001	0.069	0.1983	0.0914	0.0375	0.0004
WS4A	0.01	0.0134	0.011	0.0038	0
HPI-S2	0.005	0.0062	0.0054	0.0028	0

Table 6: Results for batch 1 for snippets in phase A of Task 4b.

System	Mean Precision	Mean Recall	Mean F-measure	MAP	GMAP
HPI-S1	0.0822	0.1706	0.0917	0.0481	0.0005
KNU-SG Team_Korea	0.0482	0.0952	0.0534	0.0266	0.0002
ustb_prir2	0.0469	0.1135	0.0503	0.0216	0.0002
ustb_prir3	0.0452	0.1070	0.0482	0.0212	0.0002
ustb_prir1	0.0409	0.1080	0.0491	0.0211	0.0002
ustb_prir4	0.0449	0.1108	0.0477	0.0201	0.0002
testtext	0.0433	0.1098	0.0460	0.0188	0.0002

Table 7: Results for batch 3 for exact answers in phase B of Task 4b.

System	Yes/no	Factoid			List		
	Accuracy	Strict Acc.	Lenient Acc.	MRR	Precision	Recall	F-measure
fa1	0.9600	0.1154	0.1923	0.1442	0.2500	0.3000	0.2641
Lab Zhu ,Fdan Univer	0.9600	0.1923	0.2692	0.2192	0.1450	0.5929	0.2181
LabZhu,FDU	0.9600	0.1923	0.2692	0.2192	0.1444	0.6214	0.2176
LabZhu_FDU	0.9600	0.1923	0.2692	0.2192	0.1420	0.5929	0.2132
Lab Zhu,Fudan Univer	0.9600	0.1923	0.2692	0.2192	0.1455	0.5770	0.2185
oaqa-3b-3	0.5200	0.2308	0.2692	0.2436	0.5396	0.5008	0.4828
WS4A	0.2400	0.0385	0.0385	0.0385	0.1172	0.2817	0.1609
LabZhu-FDU	0.0400	0.1923	0.2692	0.2192	0.1420	0.5929	0.2132

itself is very challenging. Consequently, we regard the outcome of the challenge as a success towards pushing the research on bio-medical information systems a step further. In future editions of the challenge, we aim to provide even more benchmark data derived from a community-driven acquisition process.

Acknowledgments

The fourth edition of BioASQ is supported by a conference grant from the NIH/NLM (number 1R13LM012214-01) and sponsored by the Atypon company.

References

Georgios Balikas, Ioannis Partalas, Aris Kosmopoulos, Sergios Petridis, Prodromos Malakasiotis, Ioannis Pavlopoulos, Ion Androutsopoulos, Nicolas Baskiotis, Eric Gaussier, Thierry Artieres, and Patrick Gallinari. 2013. Evaluation Framework Specifications. Project deliverable D4.1, 05/2013.

Wei CH, Leaman R, and Lu Z. 2016. Beyond accuracy: creating interoperable and scalable text-mining web services. *Bioinformatics*.

Janez Demsar. 2006. Statistical Comparisons of Classifiers over Multiple Data Sets. *Journal of Machine Learning Research*, 7:1–30.

Hyeon gu Lee, Minkyoung Kim, Juae Kim, Maengsik Choi Sunjae Kwon, Youngjoong Ko, Yi-Reun Kim, Jung-Kyu Choi, Harksoo Kim, and Jungyun Seo. 2016. KSAnswer: Question-answering System of Kangwon National University and Sogang University in the 2016 BioASQ Challenge . In *In Proceedings of the BioASQ Workshop, in ACL*.

Aris Kosmopoulos, Ioannis Partalas, Eric Gaussier, Georgios Paliouras, and Ion Androutsopoulos. 2013. Evaluation Measures for Hierarchical Classification: a unified view and novel approaches. *CoRR*, abs/1306.6802.

Chin-Yew Lin. 2004. ROUGE: A package for automatic evaluation of summaries. In *Proceedings of the ACL workshop 'Text Summarization Branches Out'*, pages 74–81, Barcelona, Spain.

James Mork, Antonio Jimeno-Yepes, and Alan Aronson. 2013. The NLM Medical Text Indexer System for Indexing Biomedical Literature. In *1st BioASQ Workshop: A challenge on large-scale biomedical semantic indexing and question answering*.

James G. Mork, Dina Demner-Fushman, Susan C. Schmidt, and Alan R. Aronson. 2014. Recent enhancements to the nlm medical text indexer. In *Proceedings of Question Answering Lab at CLEF*.

Eirini Papagiannopoulou, Yiannis Papanikolaou, Dimitris Dimitriadis, Sakis Lagopoulos, Grigorios Tsoumakas, Manos Laliotis, Nikos Markantonatos, and Ioannis Vlahavas. 2016. Large-Scale Semantic Indexing and Question Answering in Biomedicine. In *In Proceedings of the BioASQ Workshop, in ACL*.

Yannis Papanikolaou, Dimitrios Dimitriadis, Grigorios Tsoumakas, Manos Laliotis, Nikos Markantonatos, and Ioannis Vlahavas. 2014. Ensemble Approaches for Large-Scale Multi-Label Classification and Question Answering in Biomedicine. In *2nd BioASQ Workshop: A challenge on large-scale biomedical semantic indexing and question answering*.

Frederik Schulze, Ricarda Schuler, Tim Draeger, Daniel Dummer, Alexander Ernst, Pedro Flemming, Cindy Perscheid, and Mariana Neves. 2016. HPI Question Answering System in BioASQ 2016. In *In Proceedings of the BioASQ Workshop, in ACL*.

Isabel Segura-Bedmar, Adrian Carruana, and Paloma Martnez. 2016. LABDA at the 2016 BioASQ challenge task 4a: Semantic Indexing by using ElasticSearch. In *In Proceedings of the BioASQ Workshop, in ACL*.

Grigorios Tsoumakas, Ioannis Katakis, and Ioannis Vlahavas. 2010. Mining Multi-label Data. In Oded Maimon and Lior Rokach, editors, *Data Mining and Knowledge Discovery Handbook*, pages 667–685. Springer US.

Grigorios Tsoumakas, Manos Laliotis, Nikos Markontanatos, and Ioannis Vlahavas. 2013. Large-Scale Semantic Indexing of Biomedical Publications. In *1st BioASQ Workshop: A challenge on large-scale biomedical semantic indexing and question answering*.

Zi Yang, Niloy Gupta, Xiangyu Sun, Di Xu, Chi Zhang, and Eric Nyberg. 2015. Learning to answer biomedical factoid and list questions: Oaqa at bioasq 3b. In *CLEF*.

Zi Yang, Yue Zhou, and Eric Nyberg. 2016. Learning to answer biomedical questions: Oaqa at bioasq 4b. In *In Proceedings of the BioASQ Workshop, in ACL*.

Ilya Zavorin, James Mork, and Dina Demner-Fushman. 2016. Using Learning-To-Rank to Enhance NLM Medical Text Indexer Results. In *In Proceedings of the BioASQ Workshop, in ACL*.

Using Learning-To-Rank to Enhance NLM Medical Text Indexer Results

Ilya Zavorin
National Library of Medicine
Bethesda, MD, USA
Ilya.Zavorin@nih.gov

James G. Mork
National Library of Medicine
Bethesda, MD, USA
James.Mork@nih.gov

Dina Demner-Fushman
National Library of Medicine
Bethesda, MD, USA
Dina.Demner@nih.gov

Abstract

For almost 15 years, the NLM Medical Text Indexer (MTI) system has been providing assistance to NLM Indexers, Catalogers, and the History of Medicine Division (HMD) in the task of indexing the ever increasing number of MEDLINE citations, with MTI's role continuously expanding by providing more extensive and specialized coverage of the MEDLINE collection. The BioASQ Challenge has been a tremendous benefit by expanding the knowledge of leading-edge indexing research. In this paper we present an indexing approach based on the Learning to Rank methodology which was successfully applied to the indexing task by several participants of recent Challenges. The proposed solution is designed to enhance the results that come from MTI by combining strengths of MTI with additional sources of evidence to produce a more accurate list of top MeSH Heading candidates for a MEDLINE citation being indexed. It incorporates novel Learning to Rank features and other enhancements to produce performance superior to that of MTI, both overall and for two specific classes of MeSH Headings for which MTI has shown poor performance.

1 Introduction

The Indexing Section of the US National Library of Medicine® (NLM®) is tasked with processing the ever increasing number of MEDLINE®[1] citations (currently numbering more than 800,000 articles per year from more than 5,600 journals in almost 40 languages) using a vocabulary of over 27,000 MeSH® Descriptors and 220,000 MeSH Supplementary Concept Records[2]. To support this effort, various automatic and semi-automatic indexing solutions have been proposed over the years, including the NLM Medical Text Indexer (MTI) system (Mork et al., 2013).

Given any biomedical text, MTI produces a ranked list of controlled vocabulary terms (MeSH) that summarizes the main points of the text using MeSH Main Headings (MH), Subheadings (SH), Check Tags (CT), and Supplementary Concept Records (SCRs). It can also recommend a limited number of Publication Types[3] (Yepes et al., 2013a). MTI fuses heading recommendations from three separate sources: MetaMap indexing (Aronson and Lang, 2010), PubMed® Related Citations (Lin and Wilbur, 2007) and Machine Learning (Yepes et al., 2013b), with the latter source used to improve performance on some of the most frequent CheckTags. The results of this fusion are post-processed using various rules based on the end-user requirements, to provide a customized summary of the text. In this paper we focus solely on MH and CT indexing.

MTI has been made available to the research community worldwide[4] providing both a baseline for performance evaluations and input data for several other indexing systems. This includes results MTI produces for each of the weekly datasets during the BioASQ Challenges (Tsatsaronis et al., 2015).

Since 2013, the MTI team has been participating in the BioASQ Challenge which has proven to be an excellent forum for exchange and evaluation of ideas for biomedical indexing and which inspired several recent improvements in the MTI system (Mork et al., 2014). In this paper we

[1]https://www.nlm.nih.gov/pubs/factsheets/medline.html

[2]https://www.nlm.nih.gov/mesh/meshhome.html

[3]https://www.nlm.nih.gov/mesh/pubtypes.html

[4]https://ii.nlm.nih.gov/MTI/index.shtml

Proceedings of the 54th Annual Meeting of the Association for Computational Linguistics, pages 8–15,
Berlin, Germany, August 7-12, 2016.

present a component that is designed to enhance results produced by MTI. This component is based on the Learning to Rank methodology that was successfully used by several participants of recent Challenges (Liu et al., 2014a; Liu et al., 2015) . While learning from that work, we have also experimented with several new features specifically engineered to harness the power of MTI, as well as to incorporate other heterogeneous sources of evidence. We applied L2R to the results generated by MTI for the test batches of the 2016 BioASQ Challenge and other test collections comprised of recent MEDLINE citations. L2R outperformed MTI on these collections, both overall and for two specific classes of MeSH Headings for which MTI has performed poorly.

2 Learning to Rank

The task of MEDLINE indexing can be formulated as a ranking problem: given a new PubMed citation, can we find those MeSH headings that are the most relevant to this citation? In this formulation, the indexing task becomes similar to the document retrieval task, in which the documents in a collection are evaluated for relevance, significance or importance to an incoming query. In document retrieval, the documents are usually long and the queries are short, whereas in this application of ranking, the roles are in a way reversed: the citation is the query while the MeSH headings are the documents (Ruch, 2006).

In recent years, the Learning to Rank methodology (Liu, 2009) has been successfully applied to biomedical indexing. Learning to Rank (L2R) uses supervised machine learning to build a model that calculates a numerical score for any citation-heading pair. Thus, given a target citation and a set of candidate headings, L2R scores can be used to rank these candidates. The top N ranked candidates from the set are then selected as the relevant headings. The value of N is usually calculated for each citation individually.

During the training stage of L2R, a set of citations previously indexed by humans is processed to build the ranking model. For each training citation, a set of candidate headings is generated. While in principle the whole MeSH (more than 27,000 headings) may be used as candidates, in practice, only a relatively small subset of headings deemed more likely to be relevant is considered.

For each citation and each candidate heading, a feature vector is calculated. Each feature usually depends on both the citation and the heading and measures similarity between the two in some space. The features can be derived both from the raw data (such as n-grams appearing in the title and abstract of the citation and entry terms of the heading) and from metadata (such as statistics of occurrence of the heading in the journal where the citation is published). To each feature vector, a binary relevance flag is then assigned that equals 1 if the corresponding candidate MeSH heading has been assigned to the citation by a human indexer, and 0 otherwise. Assuming the same number M_T of candidates for each of the T training citations, this yields $M_T{*}T$ feature vectors with corresponding relevance flags. This training dataset is then used to build a ranking model.

Processing of a new target citation also consists of several steps. As during training, a set of candidate headings is first collected and then the corresponding set of feature vectors is generated. These vectors are ranked by the trained model and then truncated to produce the final set of recommended headings.

3 Learning-To-Rank as an MTI Booster

MTI is a mature indexing tool that provides high-accuracy recommendations for some classes of MeSH headings, such as CheckTags (Yepes et al., 2013b), while performing worse on other classes, such as "as Topic" headings[5]. It is a sophisticated multi-stage processing system that generates as output its own ranked list of candidate headings. As additional evidence, it can also produce a list of rejected candidates that, while being ultimately labeled by MTI as irrelevant based on various heuristics, have at least some relevance to the target citation. The headings at the very top of MTI ranked list for a citation are almost always correct. For example, in 2015, the percentage of correct recommendations for the highest ranked CheckTag candidates and the top five other recommendations were, respectively, 81.21%, 84.97%, 73.78%, 65.10%, 57.57%, and 51.15%, with the performance trailing off further down the list. Therefore, we choose to employ the L2R methodology to develop a complete indexing solution that uses MTI results as input. We use various types of

[5]For each "As Topic" MeSH heading, there is a corresponding Publication Type. These are designed to capture differences between what a citation is (Publication Type) versus what it is about ("as Topic" MeSH heading).

information provided by MTI to both generate the candidate heading list for a given citation and to compute some of the L2R features for these candidates. We also expand the candidate list with headings obtained from other sources, such as PubMed Related Citations (Lin and Wilbur, 2007) now known as Similar Articles, and use other types of evidence independent of MTI and PRC to generate additional features. The result is a software component that takes as input detailed MTI results for a given target citation, together with the external evidence, to produce a new list of indexing recommendations that, on average, has higher precision and recall than MTI.

Given a set of citations, each citation is processed as follows:

1. MTI is applied to the citation to produce an expanded ranked set of candidates that includes both accepted and rejected MeSH headings. For each candidate heading, we record its MTI score, whether it is a CheckTag and whether it is accepted or rejected.

2. A set of PubMed Related Citations is collected, together with their normalized similarity scores and their MeSH headings assigned by human indexers.

3. The final MeSH heading candidate set is generated as the union of MTI- and PRC-derived candidates.

4. For each candidate heading in the final set, a feature vector is calculated (see Section 4 for details).

5. In the L2R training mode, feature vectors for all citations are collected into a single training set that is used to train a model. Training is performed offline and no incremental training or tuning of the model is done afterwards.

6. In the L2R ranking mode:

   (a) The trained model computes a ranking score for each feature vector corresponding to a heading from the final candidate set.

   (b) Top candidates from the ranked list are selected as the final result (see Section 5 for more details on different ways of calculating the number of top candidates).

4 Features

4.1 PubMed Related Citations Based Features

We implemented two neighborhood features originally proposed in (Huang et al., 2011) that we denote by **PRCfreq** and **PRCsim**. They are derived from PubMed Related Citations of the citation being processed, their MeSH Headings and normalized similarity scores. For each candidate heading, **PRCfreq** is the number of PubMed Related Citations that contain this heading, and **PRCsim** is the sum of the similarity scores of those neighbors.

4.2 Text Based Features

We implemented several features that also originated from (Huang et al., 2011) and that are based on statistics collected from unigrams and bigrams extracted from the MeSH heading and its entry terms (i.e., a synonymy set of the heading) as well as the title and abstract of a citation:

- **Overlap**: The fraction of MeSH term unigrams and bigrams that appear in the title or abstract of the citation.

- **Syn**: A binary feature that captures presence of entry terms in the title and abstract.

- **IBM**: Probabilities of translating the title and abstract into a candidate MeSH heading, based on a parallel corpus of heading-title and heading-abstract pairs collected from a set of previously indexed citations and IBM statistical translation model 1 (Brown et al., 1993).

- **Okapi**: Treating the heading as the query and the title or abstract of a citation as the document, we computed similarities between the heading and the title and abstract using Okapi BM25 model (Robertson et al., 1995). Following (Mao and Lu, 2013), we used a corpus of 58,088 MEDLINE documents to construct the parallel training corpus for both Okapi and IBM features.

These features can be considered extensions of more traditional TF/IDF-based features used for ranking because TF/IDF and similar information is used for their computation. We refer the reader to (Huang et al., 2011) for further details.

4.3 Vocabulary Density Based Feature

Adding journal-specific information was shown to boost precision of MTI without losses in recall (Mork et al., 2014). We therefore included Vocabulary Density (**VocD**) as a feature in learning to rank using data provided by NLM's Indexing Initiative[6]. It is equivalent to the MeSH frequency feature described in (Liu et al., 2015).

4.4 MTI Based Features

A feature that we denote as **InMTI** is set to 1 if the candidate heading was recommended by MTI, regardless of whether or not it was included in human indexing, -1 if it was rejected by MTI and 0 otherwise. **MTIScore** is the score assigned by MTI to the corresponding candidate and divided by the score of the top MTI candidate. For PRC-derived candidates that were not recommended by MTI, this feature is set to 0. **MHtype** is a binary feature that indicates whether or not the candidate heading is a CheckTag.

4.5 Journal Descriptor Indexing Based Features

We implemented additional features based on the Journal Descriptor Indexing (JDI) methodology (Humphrey et al., 2006) maintained by the NLMs Lexical Systems Group[7]. Given a block of text, the JDI-based Text Categorization (TC) tool produces a ranked list of about 120 high-level journal descriptors (e.g. "Anatomy", "Chemistry", "Biomedical Engineering" etc) according to their relevance to the text. For example, the TC tool applied to the text "heart valve" produces ranking scores of 0.156, 0.098 and 0.090 for top three descriptors "Cardiology", "Pulmonary Medicine", and "Vascular Diseases", respectively. Similarly, JDI provides precomputed rankings of each MeSH heading against the same journal descriptors set. For example, the MeSH heading "Lung Neoplasms' has a score of 0.167 for its top descriptor "Pulmonary Medicine", 0.138 for the second closest descriptor "Neoplasms" but only 0.0187 for the descriptor "Cardiology". Given a citation text (title or abstract) and a candidate heading, we apply the TC tool to the text to find the top ranking journal descriptor, and then multiply the corresponding score by the score of the top descriptor for the heading. The more relevant the heading is to the citation text, the higher we expect the resulting product to be. We denote this feature as **JDI**.

We also implemented a simplified JDI-based feature denoted by **JDInoTC** that does not require invoking the TC tool for each heading-citation pair. Instead, it uses the journal descriptor pre-assigned to the journal where the citation is published. This assignment is designed to capture the overall topic of the journal. For example, the journal "Clinical Obesity" has been assigned the Broad Subject Term (descriptor) "Metabolism"[8]. We then set **JDInoTC** to the score of the candidate heading for that journal descriptor. Although the **JDI** and **JDInoTC** features are correlated, experiments presented in Section 5.3 show an advantage of using these features together over using just one or the other.

4.6 MeSH Similarity Based Features

We implemented a set of features inspired by the adaptation of a method called User-oriented Semantic Indexer (USI) to biomedical indexing (Fiorini et al., 2015) that uses similarity scores computed between pairs of candidate headings based on their positions in the MeSH tree, to select an optimal set of headings for a citation, without directly depending on the text of the citation. For a given candidate heading, we compute the maximum, minimum, and average MeSH-based distances from that heading to the non-rejected headings of the MTI candidate set. The intuition behind this approach is that recommending headings that are very similar to each other may be redundant while, at the other end of the distance spectrum, candidate headings that are very different from those recommended by MTI might represent spurious outliers from citations with low PRC similarity scores. The features were implemented using the SML Java library (Harispe et al., 2014). We experimented with several ways of computing pairwise heading similarity and found the combination of Jiang and Conrath semantic distance (Jiang and Conrath, 1997) with the Seco information content measure (Seco et al., 2004) to provide the best results. We denote these features as **SML**.

[6] https://ii.nlm.nih.gov/DataSets/index.shtml

[7] https://lsg2.nlm.nih.gov/LexSysGroup/Home/index.html

[8] http://www.ncbi.nlm.nih.gov/nlmcatalog/101560587

5 Experiments

We experimented with several variations of the L2R module that differed in their feature sets, their ranking algorithms, the number of PubMed Related Citations for each target citation, as well as the type of cut-off used to select the final list of recommended MeSH headings. We used the RankLib library implementation of the Learning to Rank core[9].

5.1 BioASQ 2016

To train the L2R component, as well as for local testing, we have used a dataset of 139,072 citations. This collection is comprised of randomly completed citations from the beginning of the 2015 NLM indexing year (mid-November of 2014) until early February of 2015. Since the L2R system was being actively developed at the time of the BioASQ Challenge runs, the L2R version that was evaluated had a limited number of features, namely, **PRCfreq**, **PRCsim**, **Overlap**, **Syn**, **IBM**, **Okapi**, **VocD**, and **InMTI** resulting in a feature vector of length 12. We note that in this version, unlike the one described in Section 5.3, we did not include rejected MTI candidates at either the training or the ranking stage, which also implies that the **InMTI** feature was binary. We collected 40 PubMed Related Citations for each processed citation in both training and ranking modes. When ranking a citation, we set the number of top ranked citations reported as the final result equal to the number of headings recommended by MTI. Finally, we used MART (Friedman, 2001) as the ranking algorithm. We denote this version of the L2R module applied to results of MTI as *MTI with L2R*. We also denote the default MTI system that does not use L2R as *MTI*. In Table 1 we report performance of *MTI with L2R* on two BioASQ test batches, as of May 3, 2016. Throughout this paper, we use micro-precision, recall and F_1 metrics to measure performance.

Batch/week	Precision	Recall	F_1
B 1, Wk 2	62.48%	58.81%	60.59%
B 1, Wk 3	59.09%	57.70%	58.39%
B 1, Wk 4	60.55%	54.23%	57.21%
B 1, Wk 5	58.29%	55.71%	56.97%
B 2, Wk 1	60.05%	63.26%	61.61%
B 2, Wk 1	52.74%	56.61%	54.60%
B 2, Wk 3	59.12%	55.82%	57.42%

Table 1: Performance of MTI with L2R on BioASQ 2016 Test batches 1 and 2.

5.2 Significant Improvements over MTI

We have observed that *MTI with L2R* performs significantly better than *MTI* on two specific classes of MeSH headings: Historical Check Tags and "As Topic" headings. Table 2 shows performance of *MTI with L2R* on Historical CheckTags using 2016 MTI test collection. Due to low accuracy, *MTI* currently does not recommend any Historical CheckTags except for "History, 20th Century" for which *MTI*'s precision, recall and F_1 are, respectively, 100%, 0.79%, and 1.56%. Table 3 shows performance of *MTI with L2R* for "As Topic" headings with F_1 values of at least 50%. For 39 "As Topic" headings *MTI with L2R* achieved precision of more than 50%, with 16 of those reaching perfect precision. These headings attempt to describe what an article is about (e.g. "Dissertations, Academic as Topic") whereas Publication Types attempt to capture what a citation is (e.g. "Academic Dissertations"). These differences are often subtle which leads to frequent *MTI* errors when identifying "as Topic" headings. As a result, *MTI* currently only recommends "Randomized Controlled Trials as Topic", "Patents as Topic", and "Advertising as Topic" based on a small set of trigger keywords. This yields overall precision, recall and F_1 of, respectively, 92%, 2.55% and 4.96%, which should be compared to the corresponding values from the last row of Table 3. These results demonstrate that L2R provides a significant performance boost for these two classes of MeSH headings.

Historical MH	Precision	Recall	F_1
15th Century	53.85%	28.00%	36.84%
16th Century	85.42%	73.21%	78.85%
17th Century	82.61%	51.35%	63.33%
18th Century	74.32%	55.00%	63.22%
19th Century	80.23%	64.13%	71.28%
20th Century	89.57%	70.37%	78.82%
21st Century	95.81%	26.32%	41.29%
Ancient	78.31%	51.59%	62.20%
Medieval	90.48%	66.67%	76.77%
All Historical	86.49%	54.81%	67.10%

Table 2: Performance of *MTI with L2R* on Historical CheckTags.

[9]https://sourceforge.net/p/lemur/wiki/RankLib

"As Topic" MH	Precision	Recall	F_1
D,A	100.00%	100.00%	100.00%
Cookbooks	100.00%	71.43%	83.33%
Periodicals	83.52%	63.19%	71.95%
Patents	88.89%	57.97%	70.18%
A&I	55.56%	71.43%	62.50%
W&H	83.33%	50.00%	62.50%
Formularies	66.67%	50.00%	57.14%
Poetry	85.71%	40.00%	54.55%
RS	65.38%	45.95%	53.97%
Dictionaries	100.00%	33.33%	50.00%
Manuscripts	100.00%	33.33%	50.00%
Webcasts	100.00%	33.33%	50.00%
Advertising	68.18%	39.47%	50.00%
All "as Topic"	69.56%	24.58%	36.33%

Table 3: Performance of *MTI with L2R* on individual "As Topic" headings with F_1 values of at least 50% (*"D,A", "A&I", "W&H", and "RS" denote, respectively, "Dissertations, Academic as Topic" , "Abstracting and Indexing as Topic", "Wit and Humor as Topic", and "Research Support as Topic"*), as well as collectively for all 83 "As Topic" headings.

5.3 Further L2R development

Overall, adding more features as well as using a larger number of PubMed Related Citations has a positive effect on the L2R performance. We trained L2R on the feature set from *MTI with L2R* extended with the **MHType** and **MTIScore** features and 80 PubMed Related Citations. We then experimented with other L2R configurations with additional features, and switched from MART to the LambdaMART (Wu et al., 2010) ranking method. We also compared two different ways of determining, the number of top recommendations. One approach was to preserve the number of candidates recommended by MTI (**nMTI**), as we did with *MTI with L2R*. We also observed that LambdaMART often produced positive ranking scores for the most relevant candidate headings, and negative values for irrelevant ones. Therefore the other trimming approach **PosNeg**, was to only retain the candidates with positive LambdaMART ranking scores. In some cases that produced a very long list of candidates in which case we set the threshold at 3 times the number of MTI candidates.

Table 4 shows performance of the standalone L2R module on the the 2015 MTI test collection, compared to that of MTI. It shows that **PosNeg** trimming provides a significant advantage in precision over **nMTI** with a relatively smaller drop in recall. Therefore it would be the recommended choice especially if precision is more important than recall, which is often the case during production use of the MTI system.

6 Conclusions and Future Directions

The integration of the Learning to Rank methodology as a boosting component of the MTI system improved its overall performance and showed significant gains in both precision and recall for some specific classes of MeSH headings. As is often the case in supervised machine learning, our experiments show that using a richer set of features specifically engineered to capture various types of evidence of relevance of MeSH headings to citations yields better candidate rankings. One future step in this direction would be to explore features based on author information. For example, analogous to PRC-based similarity of citations, we can explore author-based similarity. We performed limited experiments with author-derived statistics that produced some promising results. We also found that accurate author disambiguation (Liu et al., 2014b) is a prerequisite for robustness of author-based features. Other potential sources of evidence that can be used in Learning to Rank are both general and journal-specific MeSH heading coocurrence patterns[10] as well as dense distributed representations of citation text (Le and Mikolov, 2014). And to go beyond Learning to Rank, we plan to explore the application of Deep Learning to biomedical indexing and, more generally, multi-label classification (Read and Perez-Cruz, 2014).

Acknowledgments

This work was supported by the intramural research program of the U. S. National Library of Medicine, National Institutes of Health.

References

Alan R Aronson and François-Michel Lang. 2010. An overview of MetaMap: historical perspective and recent advances. *Journal of the American Medical Informatics Association*, 17(3):229–236.

Peter F Brown, Vincent J Della Pietra, Stephen A Della Pietra, and Robert L Mercer. 1993. The mathemat-

[10]https://mbr.nlm.nih.gov/MRCOC.shtml

	Precision	Recall	F_1	Precision	Recall	F_1
MTI	64.18%	63.87%	64.02%	64.18%	63.87%	64.02%
	Cut Off = nMTI			**Cut Off = PosNeg**		
L2R14F	66.88%	66.32%	66.60%	75.47%	60.79%	67.34%
L2R14F + **JDInoTC**	66.95%	66.39%	66.67%	75.66%	60.81%	67.43%
L2R14F + **JDI**	67.06%	66.49%	66.77%	75.90%	60.77%	67.49%
L2R14F + **JDInoTC** + **JDI**	67.01%	66.45%	66.73%	75.87%	60.85%	67.54%
L2R14F + **JDInoTC** + **JDI** + **SML**	67.11%	66.55%	66.83%	76.41%	60.75%	67.68%

Table 4: Performance of L2R variants on the 2015 MTI test collection. *L2R14F* model extends the original 12 features of *MTI with L2R* with the **MHType** and **MTIScore** features and 80 PubMed Related Citations.

ics of statistical machine translation: Parameter estimation. *Computational linguistics*, 19(2):263–311.

Nicolas Fiorini, Sylvie Ranwez, Sébastien Harispe, Jacky Montmain, and Vincent Ranwez. 2015. USI at BioASQ 2015: a semantic similarity-based approach for semantic indexing. In *Working Notes for the Conference and Labs of the Evaluation Forum (CLEF), Toulouse, France.*

Jerome H Friedman. 2001. Greedy function approximation: a gradient boosting machine. *Annals of statistics*, pages 1189–1232.

Sébastien Harispe, Sylvie Ranwez, Stefan Janaqi, and Jacky Montmain. 2014. The semantic measures library and toolkit: fast computation of semantic similarity and relatedness using biomedical ontologies. *Bioinformatics*, 30(5):740–742.

Minlie Huang, Aurélie Névéol, and Zhiyong Lu. 2011. Recommending MeSH terms for annotating biomedical articles. *Journal of the American Medical Informatics Association*, 18(5):660–667.

Susanne M Humphrey, Willie J Rogers, Halil Kilicoglu, Dina Demner-Fushman, and Thomas C Rindflesch. 2006. Word sense disambiguation by selecting the best semantic type based on Journal Descriptor Indexing: Preliminary experiment. *Journal of the American Society for Information Science and Technology*, 57(1):96–113.

Jay J Jiang and David W Conrath. 1997. Semantic similarity based on corpus statistics and lexical taxonomy. *arXiv preprint cmp-lg/9709008*.

Quoc Le and Tomas Mikolov. 2014. Distributed representations of sentences and documents. In *Proceedings of the 31st International Conference on Machine Learning (ICML-14)*, pages 1188–1196.

Jimmy Lin and W John Wilbur. 2007. PubMed related articles: a probabilistic topic-based model for content similarity. *BMC bioinformatics*, 8(1):423.

Ke Liu, Junqiu Wu, Shengwen Peng, Chengxiang Zhai, and Shanfeng Zhu. 2014a. The fudan-uiuc participation in the BioASQ challenge task 2a: The antinomyra system. *CLEF2014 Working Notes*, 129816:100.

Wanli Liu, Rezarta Islamaj Doğan, Sun Kim, Donald C Comeau, Won Kim, Lana Yeganova, Zhiyong Lu, and W John Wilbur. 2014b. Author name disambiguation for PubMed. *Journal of the Association for Information Science and Technology*, 65(4):765–781.

Ke Liu, Shengwen Peng, Junqiu Wu, Chengxiang Zhai, Hiroshi Mamitsuka, and Shanfeng Zhu. 2015. MeSHLabeler: improving the accuracy of large-scale mesh indexing by integrating diverse evidence. *Bioinformatics*, 31(12):i339–i347.

Tie-Yan Liu. 2009. Learning to rank for information retrieval. *Foundations and Trends in Information Retrieval*, 3(3):225–331.

Yuqing Mao and Zhiyong Lu. 2013. NCBI at the 2013 BioASQ challenge task: Learning to rank for automatic mesh indexing. Technical report, Technical report.

James G Mork, Antonio Jimeno-Yepes, and Alan R Aronson. 2013. The NLM Medical Text Indexer system for indexing biomedical literature. In *BioASQ@ CLEF.*

James G Mork, Dina Demner-Fushman, Susan Schmidt, and Alan R Aronson. 2014. Recent enhancements to the NLM Medical Text Indexer. In *CLEF (Working Notes)*, pages 1328–1336.

Jesse Read and Fernando Perez-Cruz. 2014. Deep learning for multi-label classification. *arXiv preprint arXiv:1502.05988.*

Stephen E Robertson, Steve Walker, Susan Jones, Micheline M Hancock-Beaulieu, Mike Gatford, et al. 1995. Okapi at TREC-3. *NIST SPECIAL PUBLICATION SP*, 109:109.

Patrick Ruch. 2006. Automatic assignment of biomedical categories: toward a generic approach. *Bioinformatics*, 22(6):658–664.

Nuno Seco, Tony Veale, and Jer Hayes. 2004. An intrinsic information content metric for semantic similarity in wordnet. In *ECAI*, volume 16, page 1089.

George Tsatsaronis, Georgios Balikas, Prodromos Malakasiotis, Ioannis Partalas, Matthias Zschunke, Michael R Alvers, Dirk Weissenborn, Anastasia Krithara, Sergios Petridis, Dimitris Polychronopoulos, et al. 2015. An overview of the bioasq large-scale biomedical semantic indexing and question answering competition. *BMC bioinformatics*, 16(1):1.

Qiang Wu, Christopher JC Burges, Krysta M Svore, and Jianfeng Gao. 2010. Adapting boosting for information retrieval measures. *Information Retrieval*, 13(3):254–270.

Antonio J Jimeno Yepes, James G Mork, and Alan R Aronson. 2013a. Identifying publication types using machine learning.

Antonio Jose Jimeno Yepes, James G Mork, Dina Demner-Fushman, and Alan R Aronson. 2013b. Comparison and combination of several MeSH indexing approaches. In *AMIA annual symposium proceedings*, volume 2013, page 709. American Medical Informatics Association.

LABDA at the 2016 BioASQ challenge task 4a: Semantic Indexing by using ElasticSearch

Isabel Segura-Bedmar, Adrián Carruana, Paloma Martínez
Computer Science Department, University Carlos III of Madrid
Avd. Universidad, 30, Leganés, 28911, Madrid, Spain
isegura,acarruan,pmf@inf.uc3m.es

Abstract

This paper describes the participation of LABDA team in the 2016 BioASQ Task 4a on large-scale online biomedical semantic indexing. Our approach is based on the use of the open source search engine ElasticSearch. Experimental results show that our approach achieves high recall while keeping processing time low. Although more work needs to be done to improve our results, we can conclude that ElasticSearch is a competitive and scalable system for indexing biomedical literature.

1 Introduction

Biomedical Natural Language Processing (BioNLP) has made great advances in the last decade thanks to different community-wide challenge evaluations, such as BioCreative (Krallinger et al., 2015), BioNLP shared tasks (Kim et al., 2011; Nédellec et al., 2013), i2b2 (Stubbs et al., 2015) DDIExtraction (Segura-Bedmar et al., 2011; Segura Bedmar et al., 2013), etc. While most of them have pursued the further development of research on informations extraction tasks, the BioASQ Challenge[1] focuses on biomedical semantic indexing and question answering fields.

Biomedical Semantic Indexing is to identify the MeSH categories that best describe a PubMed article and is a crucial task to facilitate literature search. This process is manually performed by human experts, thus becoming a costly, time-consuming and laborious task (Huang et al., 2011). Therefore there is an urgent need to explore automatic methods to support this task.

As in previous editions (Tsatsaronis et al., 2015; Balikas et al., 2015), BioASQ 2016 consists of two different tasks: large-scale online biomedical semantic indexing (Task 4a) and question answering (Task 4b). This paper describes our participation in Task 4a. The goal of the task is to automatically predict the most relevant MeSH labels for a given document. One of the major challenges of the task is to manage scalability due to the great amount of documents that have to be indexed. More than 750,000 articles were added in 2014 with a load of 2000-4000 documents per day.[2] Search systems such as ElasticSearch, an open source search engine, could be adequate frameworks to cope with this information overload problem.

To the best of our knowledge, this is the first work that addresses semantic indexing by using ElasticSearch. Due to the horizontal scalability provided by ElasticSearch, it is possible to index large collections of documents, as is the case of the Medline/PubMed database with more than 22 million citations to date. Our approach is to index the training set provided by the BioASQ organizers with ElasticSearch. Then, each document in the test set is translated into a query, that is fired against the index built from the training set, returning the most relevant documents and their MeSH categories. Finally, each MeSH category is ranked using a scoring system based on the frequency of the category and the similarity of relevant documents, which contain the category, with the test document to classify. Up to date at which we write this paper, no official definitive results have been published for any of our submissions yet. To evaluate our approach, we generated our own development set from a random sample of 1099 training documents. To avoid any potential bias, these documents were removed from the training set. Tested on this development set, our approach achieves a recall of 80.6%, precision of 45.4% and an F1 of

[1] http://www.bioasq.org/

[2] https://www.nlm.nih.gov/pubs/factsheets/medline.html

Proceedings of the 54th Annual Meeting of the Association for Computational Linguistics, pages 16–22,
Berlin, Germany, August 7-12, 2016.

56.3%. In comparison to the Medical Text Indexer (MTI) (Mork et al., 2013), which is considered the baseline system of the task, our system does not only provide an improvement of more than 1% in F1, but also has a much better time response (15 seconds per document) than the MTI system (30-45 seconds per document).[3]

The rest of the paper is organized as follows: related work is presented in Section 2. Section 3 presents a description of our method and the datasets used in this study. Then, we report and discuss some preliminary results of our approach in section 4. Finally, section 5 presents conclusion and future work.

2 Related Work

Semantic indexing of MEDLINE articles is a manual laborious task which could be helped by information technology. The objective is to tag an article with a set of MeSH categories, hence it is a multilabel classification problem.

The main challenge of this shared task is to work with MeSH, a big hierarchy that includes a controlled vocabulary composed of 15 root concepts, such as organisms and diseases, with more than 25,000 categories. Most of works restrict the scope of MeSH hierarchy using only a particular branch in the MeSH tree (for instance Heart Diseases) (Ruiz and Srinivasan, 2002), or a subset of tags, generally those appearing in the training collection (Yepes et al., 2015).

Current state-of-the art includes approaches whose general architecture comprises two differentiated phases: a first phase that obtains an initial set of MeSH categories that could represent the document to classify and a second phase that rerank these categories to select the top K that better fit the input document. In both phases different document features can be used; the most frequent feature model is the so called bag-of-words (where words could follow a ngram model or be a word, phrase, concept, etc. storing a value that represents its presence frequency in the document or any other model such as TF*IDF).

Doing a review of BioASQ previous editions (Partalas et al., 2013; Balikas et al., 2014; Balikas et al., 2015; Tsatsaronis et al., 2015), the main characteristics of participants systems are: approaches that use flat methods which consider each MeSH category independently of the others or hierarchical methods that take into account the MeSH tree structure; the machine learning techniques used to select the initial set of MeSH labels (SVM, logistic regression, K nearest neighbor, etc.); the word model (unigram, bigram, trigram); if Natural Language Processing (NLP) tools are included to preprocess documents (POS taggers, chunkers, syntactic parser); if domain specific resources are used (for instance, UMLS ontology or WordNet lexical database); if the system is built over a search-based platform (such as Lucene); if curator annotation guidelines are considered and the processing and storage requirements both in the definition of models to multilabel training and classification process.

In 2013 edition, the best systems (depending on the batch) were the Medical Text Indexer (MTI) (Mork et al., 2013) with a micro F measure of 0.5481 and the system AUTH (Tsoumakas et al., 2013) with a micro F measure of 0.578. The MTI system, which is considered the baseline system of the task, is based on a combination of Metamap indexing and Pubmed related citations to recognize MeSH concepts that then are clustered and ranked. AUTH system preprocessed the articles using the Stanford parser and bigram frequencies were extracted. The meta-labeler tool (Tang et al., 2009), which is based on SVM binary classifiers trained for each label present in a subset of training collection, was used to rank the labels and a regression model is used to predict the K top labels.

In 2014 edition, several systems outperformed the MTI baseline system (micro F measure of 0.547), the system of NCBI (Mao et al., 2014), with a micro F measure of 0.605 and the Antinomyra system (Liu et al., 2014), with a micro F measure of 0.619. The NCBI system selected the relevant MeSH labels for a given article from its k-nearest neighbor documents. This set was also extended with the MeSH labels proposed by the MTI system. Then, a learning-to-rank algorithm was used to sort the MeSH labels based on the learned associations between the article text and each MeSH label. This system also used SVM binary classifiers (trained for each MeSH label in the training data) to predict the MeSH labels in the test data. The Antinomyra system followed a similar approach but instead of using SVM classifiers it used a logistic regression method.

[3]https://www.nlm.nih.gov/mesh/MeSHonDemand.html

The winner in BioASQ 2015 (Liu et al., 2015) used a learning to rank approach that returns an ordered list of MeSH categories for each instance using a combination of binary classifiers, similar articles to the article to annotate, pattern matching between MeSH categories and title of the article as well as the prediction of the MTI baseline system. This system achieved a micro F measure of 0.615.

Concerning the annotation guidelines followed by curators, some works such as (Mork et al., 2013) make use of MEDLINE annotation guidelines to postprocess the ranking of MeSH categories. The overview of BioASQ 2013 systems (Tsatsaronis et al., 2015) suggests that it is difficult to know the utility of the is-a relations in the MeSH hierarchy due to human curators do not seem to follow the annotation guidelines concerning the use of most specialized tag.

Out of the scope of BioASQ forum, the approach described in (Rak et al., 2007), was based on association rule mining from the OHSUMED corpus (Hersh et al., 1994), which contains approximately 340.000 articles from 1987 to 1991 (the rules are a kind of information retrieval techniques where a set of words determine the class of the document). A more recent work (Yepes et al., 2015) analyzed different representations of articles based on lexical, syntactic and semantic information. This system was tested over a collection of 143,853 citations and 63 selected MeSH categories (those with at least 1,500 citations indexed). Application of NLP features do not exhibit good performance although combination of all features performs better than individual sets. Participants in BioASQ such as (Ribadas et al., 2014) achieved poor results when NLP techniques are included.

3 Method

The goal of the task is to automatically predict the most relevant MeSH categories for each article in a test set. The predictions should be compared to MeSH categories proposed by human curators. This section describes the method and data used in this study.

3.1 Data

The training data for the BioASQ task 4.a consist of PubMed articles that were manually annotated with MeSH terms by human curators. In addition to the new 2016 training dataset, the training datasets of the previous BioASQ challenges are available too. The main difference between those datasets is the version of the MeSH vocabulary that was used to annotate their articles. It should be noted that each year a new release of MeSH including updates of its structure (for example, 310 new MeSH Headings were added to MeSH in 2015) is published. Typically, articles are not re-indexed with the new MeSH terms.

The teams are permitted to use any resource to train their systems, however we only use the 2016 training dataset because the evaluation will be performed using the MeSH version 2016. There are two versions of the training data: (1) Training v.2016a with more than 12 million of documents, and (2) Training v.2016b with almost 5 million of documents from the pool of journals that the BioASQ organizers use to select the articles for the test data. This dataset was built using only journals with small average annotation periods. In both datasets, the average number of MeSH terms assigned to an article is 12-13.

In order for the teams to evaluate their systems, a new test set is available every Monday. Then, the teams can upload their results before the next 24 hours after the release. A total 15 test sets have been published, which are grouped in three different periods (batches). It should be noted that the articles used in the test datasets have not been annotated yet by human experts, and therefore, it is not possible to provide an immediate evaluation of the participant systems. This is an important inconvenience since there is no fast way to assess if a given technique or resource helps to improve the results. It should be very helpful having a development dataset. We built our own development dataset from a random sample of 1099 documents taken from the small training dataset (Training v.2016a). Thus, our development dataset only collects articles from the same set of journals used to build the test datasets of the task. As mentioned above, these articles were removed from our training set in order to avoid any potential bias.

3.2 ElasticSearch

Our approach relies on the assumption that similar documents should be classified by similar MeSH labels. While previous work has exploited a kNN approach in order to propose the MeSH labels of the relevant documents for a given query (test document), we propose to calculate document similarity by using ElasticSearch, an open source search engine. ElasticSearch provides horizontal scalability, that is, it is able to index large collections of documents. The main advantage of ElasticSearch is its capacity to create distributed systems by specifying only the configuration of the hierarchy of nodes. Then, ElasticSearch is self-managed to maintain better fault tolerance and load distribution. The core of ElasticSearch is Lucene,[4] a free, open-source and de-facto standard retrieval software library. Lucene is based on the well-known and commonly used vector space model for information retrieval. The efficiency of Lucene is due to it searches on index instead of searching the text directly.

Another important advantage of ElasticSearch is that it does not require very high computing power and a high storage capacity to index large collections. In this study, ElasticSearch (version 2.2) was installed on a server Ubuntu Server 14.0f4 with 24GB of RAM and 500GB of disk space. We create an index (that is like a database in a relational database) built from the training dataset. By default, each index in Elasticsearch is configured with five shards, lucene instances. One of the most important advantages of ElasticSearch is that the shards can be distributed amongst all nodes in the cluster, and can be moved from one node to another in the case of node failure. Each shard has a backup copy.

As it has already been mentioned before, our approach is to index the training dataset and represent each test document as a query. In particular, we define two different types of index, one using the large training dataset (Training v.2016a) and the second one using the small training set (Training v.2016b), that only contains articles from the journals used for testing. Both collections are indexed using bag-of-word model. To translate the test documents to the queries, each document is also represented as bag of words. Then, each query is fired against the index, returning the most relevant documents (relevance scoring is calculated using TF/IDF). Figure 1 shows the basic architecture of our system.

Finally, the MeSH categories from the relevant documents are collected. The simplest approach would be to return the whole set of MeSH labels for all retrieved documents. However, we define a metric to rank each MeSH category for a given test document based on the total number of occurrences of the label in the whole index as well as the similarity of the relevant document containing this category with the test document (query). Our scoring system is based on the hypothesis that similar documents should have similar MeSH categories, and that the most used MeSH categories should achieve higher scores. The following formula describes this metric:

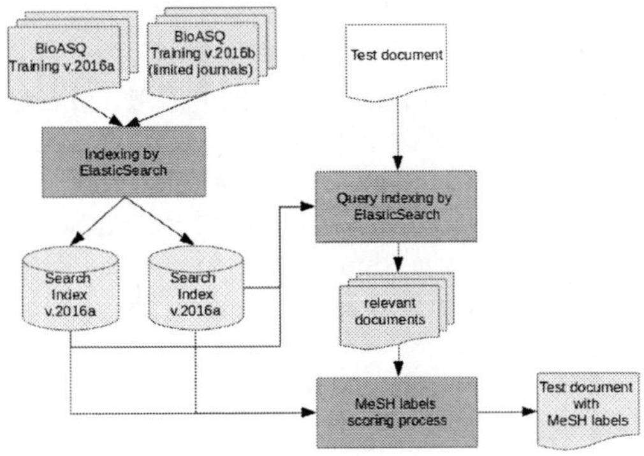

Figure 1: Architecture of our system.

$$score(l,q) = (tf(l) - R) \sum_{d:l \in d} score(d,q) \quad (1)$$

where $tf(l)$ refers to the total number of occurrences of the label in the whole index, and R is a discrete parameter that indicates the minimum number (minus one) of times a label has to appear in the relevant documents in order to be considered as a candidate label for the test document. R takes only three values: 0, 1 and 2. Finally, $score(d, q)$ represents the scoring of a document *d*, containing the MeSH label *l*, for a given query *q* (a test document).

While some documents present a large number of MeSH labels, others only contain a small set. In order to reduce this variability, the scoring for a label is normalized using the following equation:

$$score(l,q)_n = (tf(l) - R) \sum_{d:l \in d} \frac{score(d,q)}{max_{a:l \in a} score(a,q)} \quad (2)$$

Finally, we choose those MeSH categories with a score higher than a threshold (which was set empirically upon our development dataset). It should be noted that if the threshold is set to 0 then the whole set of MeSH categories for all retrieved documents is returned.

[4]https://lucene.apache.org/core/

4 Experimental results

Task 4.a began on 8th of February, 2016, however we enrolled almost two months later. Our first submission was on the fourth week of the second batch (March 14-April 11). Unfortunately, there is no results for our systems at the time of writing this paper and we cannot offer any official definitive results. For this reason, we show the results of our settings on own development dataset.

The performance of the participating systems is evaluated using standard IR measures (e.g., precision, recall, accuracy), as well as hierarchical variants of them, such as Lowest Common Ancestor F-measure (LCA-F). The HEMKit tool[5] (Kosmopoulos et al., 2015) was used to evaluate our different settings on our development set.

We experimented with different settings such as the index used to retrieve the documents, the number of relevant documents (10, 20, 30 and 40), the option of including MeSH labels without repetitions, and the threshold to select the MeSH labels. We also provided baseline results based on the use of the MTI system (Mork et al., 2013). Table 1 shows the results. Our best result among all approaches is highlighted in bold. The different settings are described bellow:

- **MTI**: our baseline system using MTI.
- **Elastic-2016V-X-R-T**: V refers to the index used: a for the index built from the large training dataset (Training v.2016a) or b for index built from the small training dataset (Training v.2016b). X refers to the number of relevant documents retrieved by ElasticSearch. R is a discrete parameter that indicates the minimum number (minus one) of times a label has to appear in the relevant documents in order to be considered as a candidate label for the test document. R takes only three values: 0, 1 and 2. T refers to the minimum threshold in equation 2 for selecting the MeSH labels.

We experimented with different settings such as the index type, the number of relevant documents or the threshold used to select the MeSH categories. Results for some of these settings are shown in Table 1 (we do not show all results for lack of space). Experiments showed that the increase in the number of relevant documents achieved to improve precision and recall values. Finally, the number of documents was set to 30 because this value achieved the best results while keeping the processing time low (less than 15 seconds per document).

The simplest approach by using ElasticSearch (that is, returning the whole set of MeSH labels for all retrieved documents) provides a very high recall (93%) but with a very low precision (15-16%). We tried with different values for the threshold T (minimum score to select the MeSH categories) and decided that 1.5 was a good value balancing precision with recall as higher values returned.

Regardless of the other parameters, the index type, that is, the use of the large training dataset versus the small training dataset, does not seem to obtain a significant difference. The results obtained with the small index are slightly better than those obtained with the large index.

As could be expected, the fact of including the MeSH categories with frequencies lower than 2 achieves better recall value, but has worse precision. On the contrary, if we require that the MeSH category has to occur at least twice in the set of the relevant documents in order to be selected, the precision increases but the recall decreases.

When comparing the experimental results of the current study with those from the MTI baseline, we can observe that our approach outperforms this baseline at recall, but with a significant decrease in precision. Therefore, we need to further research for techniques to improve precision. On the other hand, it should be noted that our system based on ElasticSearch gives a much better time response than the MTI system.

Finally, we also combined the MTI baseline with our approach based on ElasticSearch by outputing all MeSH labels proposed by MTI as well as those proposed by ElasticSearch. In this case, the best value for the threshold T was 3. This setting provided the best results (see two last rows in Table 1).

Table 2 shows our results on a very small sample (302 articles) from the test batch 3-week 5, and thereby, no conclusion can be drawn yet. The setting used for this submission was only based on providing the labels from the top 30 articles retrieved by ElasticSearch from the small index (Training v.2016b). This set was also extended with the MeSH labels proposed by MTI.

[5]http://nlp.cs.aueb.gr/software

Systems	F	R	P	LCA-F	LCA-R	LCA-P
MTI	0.7065	0.6741	0.6881	0.4165	0.4217	0.454
Elastic-2016a-30-0-0	0.2734	0.9394	0.1647	0.2004	0.6792	0.1206
Elastic-2016b-30-0-0	0.2626	0.9364	0.1571	0.1933	0.6752	0.1156
Elastic-2016a-30-1-1.5	0.5150	0.8303	0.3926	0.3345	0.5589	0.2510
Elastic-2016b-30-1-1.5	0.5188	0.8474	0.3925	0.3377	0.5717	0.2519
Elastic-2016a-30-2-1.5	0.5592	0.7944	0.4537	0.3580	0.5282	0.2861
Elastic-2016b-30-2-1.5	0.5632	0.8066	0.4543	0.3625	0.5364	0.2889
MTI + Elastic-2016a-30-2-3	**0.6266**	0.8168	0.5330	0.4034	0.5420	0.3396
MTI + Elastic-2016b-30-2-3	**0.6207**	0.8039	0.5297	0.3982	0.5345	0.3357

Table 1: Experimental results on our development dataset.

Systems	F	P	R	LCA-F	LCA-P	LCA-R
MTI	0.6373	0.6650	0.6674	0.3949	0.4085	0.4168
MTI + Elastic-2016b-30-2-3	0.4408	0.3295	0.6910	0.3890	0.4774	0.7928

Table 2: Experimental results on the test batch 3, week 5 (Annotated articles:302/3130).

5 Conclusions

Several works have already applied a k-Nearest-Neighbors (kNN) approach for semantic indexing (Névéol et al., 2007; Mao et al., 2014; Dramé et al., 2014). This approach relies on the assumption that similar documents should be classified by similar MeSH labels. We make the same assumption, but our work is the first that explores the document similarity using ElasticSerach instead of kNN. Our approach achieves similar results to those reported in previous editions of BioASQ, while keeping the processing time much lower than that reported by the MTI baseline (30-45 seconds per document). Our approach yields high recall (80-84%), but with a low precision (45-53%). Therefore, we plan to study alternatives that aim to improve precision. As future steps, we also plan to determine semantic similarity between documents using word embeddings (Mikolov et al., 2013), instead of the well-known and commonly used vector space model for information retrieval.

Acknowledgments

This work was supported by eGovernAbility-Access project (TIN2014-52665-C2-2-R).

References

George Balikas, Ioannis Partalas, Axel-Cyrille Ngonga Ngomo, Anastasia Krithara, Eric Gaussier, and Georgios Paliouras. 2014. Results of the BioASQ tasks of the Question Answering Lab at CLEF 2014. In *Proceedings of CLEF 2014*.

Georgios Balikas, Aris Kosmopoulos, Anastasia Krithara, Georgios Paliouras, and Ioannis Kakadiaris. 2015. Results of the BioASQ tasks of the Question Answering Lab at CLEF 2015. In *Proceedings of CLEF 2015*.

Khadim Dramé, Fleur Mougin, and Gayo Diallo. 2014. A k-nearest neighbor based method for improving large scale biomedical document indexing. In *Proceedings of the 6th International Symposium on Semantic Mining in Biomedicine (SMBM)*, pages 19–26.

William Hersh, Chris Buckley, TJ Leone, and David Hickam. 1994. OHSUMED: An interactive retrieval evaluation and new large test collection for research. In *Proceedings of SIGIR94*, pages 192–201.

Minlie Huang, Aurélie Névéol, and Zhiyong Lu. 2011. Recommending MeSH terms for annotating biomedical articles. *Journal of the American Medical Informatics Association*, 18(5):660–667.

Jin-Dong Kim, Sampo Pyysalo, Tomoko Ohta, Robert Bossy, Ngan Nguyen, and Jun'ichi Tsujii. 2011. Overview of BioNLP shared task 2011. In *Proceedings of the BioNLP Shared Task 2011 Workshop*, pages 1–6.

Aris Kosmopoulos, Ioannis Partalas, Eric Gaussier, Georgios Paliouras, and Ion Androutsopoulos. 2015. Evaluation measures for hierarchical classification: a unified view and novel approaches. *Data Mining and Knowledge Discovery*, 29(3):820–865.

Martin Krallinger, Obdulia Rabal, Florian Leitner, Miguel Vazquez, David Salgado, Zhiyong Lu,

Robert Leaman, Yanan Lu, Donghong Ji, Daniel M Lowe, et al. 2015. The CHEMDNER corpus of chemicals and drugs and its annotation principles. *Journal of Cheminformatics*, 7(S1):1–17.

Ke Liu, Junqiu Wu, Shengwen Peng, Chengxiang Zhai, and Shanfeng Zhu. 2014. The FUDAN-UIUC participation in the BioASQ challenge Task 2a: The antinomyra system. In *CLEF2014 (Working Notes)*, volume 129816, page 100.

Ke Liu, Shengwen Peng, Junqiu Wu, Chengxiang Zhai, Hiroshi Mamitsuka, and Shanfeng Zhu. 2015. MeSHLabeler: improving the accuracy of large-scale MeSH indexing by integrating diverse evidence. *Bioinformatics*, 31(12):i339–i347.

Yuqing Mao, Chih-Hsuan Wei, and Zhiyong Lu. 2014. NCBI at the 2014 BioASQ Challenge Task: Large-scale Biomedical Semantic Indexing and Question Answering. In *CLEF2014 (Working Notes)*, pages 1319–1327.

Tomas Mikolov, Ilya Sutskever, Kai Chen, Greg S Corrado, and Jeff Dean. 2013. Distributed representations of words and phrases and their compositionality. In *Advances in neural information processing systems*, pages 3111–3119.

James G Mork, Antonio Jimeno-Yepes, and Alan R Aronson. 2013. The NLM Medical Text Indexer System for Indexing Biomedical Literature. In *Proceedings of BioASQ CLEF 2013*.

Claire Nédellec, Robert Bossy, Jin-Dong Kim, Jung-Jae Kim, Tomoko Ohta, Sampo Pyysalo, and Pierre Zweigenbaum. 2013. Overview of BioNLP shared task 2013. In *Proceedings of the BioNLP Shared Task 2013 Workshop*, pages 1–7.

Aurélie Névéol, James G Mork, and Alan R Aronson. 2007. Automatic indexing of specialized documents: using generic vs. domain-specific document representations. In *Proceedings of the Workshop on BioNLP 2007: Biological, Translational, and Clinical Language Processing*, pages 183–190.

Ioannis Partalas, Éric Gaussier, and Axel-Cyrille Ngonga Ngomo. 2013. Results of the First BioASQ Workshop. In *Proceedings of BioASQ CLEF 2013*, pages 1–8.

Rafal Rak, Lukasz A Kurgan, and Marek Reformat. 2007. Multilabel associative classification categorization of MEDLINE articles into MeSH keywords. *IEEE engineering in medicine and biology magazine*, 26(2):47.

Francisco J Ribadas, Luis M De Campos, Vıctor M Darriba, and Alfonso E Romero. 2014. CoLe and UTAI participation at the 2014 BioASQ semantic indexing challenge. In *Proceedings of the CLEF BioASQ 2014 Workshop*, pages 1361–1374.

Miguel E Ruiz and Padmini Srinivasan. 2002. Hierarchical text categorization using neural networks. *Information Retrieval*, 5(1):87–118.

Isabel Segura-Bedmar, Paloma Martinez, and Daniel Sanchez-Cisneros. 2011. The 1st DDIExtraction-2011 Challenge Task: Extraction of Drug-Drug Interactions from Biomedical Texts. In *Proceedings of the 1st Challenge Task on Drug-Drug Interaction Extraction 2011*, pages 1–9.

Isabel Segura Bedmar, Paloma Martínez, and María Herrero Zazo. 2013. Semeval-2013 task 9: Extraction of drug-drug interactions from biomedical texts (ddiextraction 2013). In *Second Joint Conference on Lexical and Computational Semantics (*SEM)*, volume 2: Seventh International Workshop on Semantic Evaluation (SemEval 2013), page 341350.

Amber Stubbs, Christopher Kotfila, Hua Xu, and Özlem Uzuner. 2015. Identifying risk factors for heart disease over time: overview of 2014 i2b2/UTHealth shared task Track 2. *Journal of Biomedical Informatics*, 58:S67–S77.

Lei Tang, Suju Rajan, and Vijay K Narayanan. 2009. Large scale multi-label classification via metalabeler. In *Proceedings of the 18th international conference on World Wide Web*, pages 211–220.

George Tsatsaronis, Georgios Balikas, Prodromos Malakasiotis, Ioannis Partalas, Matthias Zschunke, Michael R Alvers, Dirk Weissenborn, Anastasia Krithara, Sergios Petridis, Dimitris Polychronopoulos, et al. 2015. An overview of the BioASQ large-scale biomedical semantic indexing and question answering competition. *BMC Bioinformatics*, 16(1):1.

Grigorios Tsoumakas, Manos Laliotis, Nikos Markantonatos, and Ioannis Vlahavas. 2013. Large-scale semantic indexing of biomedical publications at bioasq. In *Proceedings of BioASQ CLEF 2013*.

Antonio Jose Jimeno Yepes, Laura Plaza, Jorge Carrillo-de Albornoz, James G Mork, and Alan R Aronson. 2015. Feature engineering for MEDLINE citation categorization with MeSH. *BMC Bioinformatics*, 16(1):1.

Learning to Answer Biomedical Questions: OAQA at BioASQ 4B

Zi Yang Yue Zhou Eric Nyberg
Language Technologies Institute
School of Computer Science
Carnegie Mellon University
{ziy, ehn}@cs.cmu.edu

Abstract

This paper describes the OAQA system evaluated in the BioASQ 4B Question Answering track. The system extends the Yang et al. (2015) system and integrates additional biomedical and general-purpose NLP annotators, machine learning modules for search result scoring, collective answer reranking, and yes/no answer prediction. We first present the overall architecture of the system, and then focus on describing the main extensions to the Yang et al. (2015) approach. Before the official evaluation, we used the development dataset (excluding the 3B Batch 5 subset) for training. We present initial evaluation results on a subset of the development data set to demonstrate the effectiveness of the proposed new methods, and focus on performance analysis of yes/no question answering.

1 Introduction

The BioASQ QA challenge (Tsatsaronis et al., 2015) evaluates automatic question answering technologies and systems in the biomedical domain. It consists of two phases: in Phase A, the task requires to retrieve relevant document, snippets, concepts, and triples given a natural language question, and evaluates the retrieval results in terms of mean average precision (MAP); in Phase B, the task requires to generate ideal answers for the questions, which are evaluated using accuracy and mean reciprocal rank (MRR), as well as exact answers, which are evaluated based on manual judgment. The OAQA team participated in Batches 3, 4, and 5 of BioASQ 4B, in the categories of document, snippet, and concept retrieval, factoid, list and yes/no question answering (exact answer generation). The source code of the participating system can be downloaded from our GitHub repository[1].

We follow the same general hypothesis expressed in Ferrucci et al. (2009) and Yang et al. (2015), specifically that informatics challenges like BioASQ are best met through careful design of a flexible and extensible architecture, coupled with continuous, incremental experimentation and optimization over various combinations of existing state-of-the-art components, rather than relying on a single "magic" component or single component combination. This year, the number of labeled questions in the development set has grown to 1,307 (up from 810 in last year's dataset), which allows further exploration of a) the potential of supervised learning methods, and b) the effectiveness of various biomedical NLP tools in various phases of the system, from relevant concept and document retrieval to snippet extraction, and from answer text identification to answer prediction.

First, we use TmTool[2] (Wei et al., 2016), in addition to MetaMap[3], to identify possible biomedical named entities, especially out-of-vocabulary concepts. We also extract frequent multi-word terms from relevant snippets (Frantzi et al., 2000) to further improve the recall of concept and candidate answer text extraction. Second, we propose a supervised learning method to rerank the answer candidates for factoid and list questions based on the relation between each candidate answer and other candidate answers, which we refer to as collective reranking in this paper. Third, we implement a yes/no question answering pipeline combining various heuristics, e.g. negation words, sentiment of the statements, the biomedical con-

[1] `https://github.com/oaqa/bioasq`

[2] `http://www.ncbi.nlm.nih.gov/CBBresearch/Lu/Demo/tmTools/`

[3] `http://metamap.nlm.nih.gov/`

Proceedings of the 54th Annual Meeting of the Association for Computational Linguistics, pages 23–37,
Berlin, Germany, August 7-12, 2016.

cepts mentioned in the relevant snippets that belong to the same concept type, and question inversion (Kanayama et al., 2012). Finally, we introduce a unified classification interface for judging the relevance of each retrieved concept, document, and snippet, which can combine the relevant scores evidenced by various sources, e.g. retrieval scores using different queries and indexes.

This paper describes the system that was evaluated in the BioASQ 4B challenge. We first review the system architecture and the approaches used in Yang et al. (2015) in Section 2, and then we focus on describing each individual component for BioASQ 4B in Sections 3 to 6. Before the official evaluation, we trained the system using the development dataset excluding the 3B Batch 5 subset; we evaluate the proposed approach using the held-out 3B Batch 5 subset. Section 7 presents the results, which illustrate the effectiveness of the proposed methods, and Section 8 presents a manual error analysis of the proposed yes/no QA method and highlight the challenges of biomedical yes/no QA problem. We conclude and present future work in Section 9.

2 Overview of Yang et al. (2015) System

In this section, we briefly describe the architecture of the Yang et al. (2015) system, which provided the baseline for the system evaluated here. Further detail can be found in the full paper.

The Yang et al. (2015) system uses the UIMA ECD/CSE framework[4] (Garduno et al., 2013; Yang et al., 2013) with a YAML[5]-based language to support formal, declarative descriptors for the space of system and component configurations to be explored during system optimization. The system employs a three-layered architecture. The first layer BaseQA[6] is designed for domain-independent QA components, and includes the basic input/output definition of a QA pipeline, intermediate data objects, QA evaluation components, and data processing components. In the second layer, we implemented biomedical resources that can be used in any biomedical QA task (outside the context of BioASQ). A few BioASQ-specific components were integrated in the third design layer; for example, GoPubMed services are only hosted for the purpose of the BioASQ challenge. Tools and resources that are shared by multiple components are defined as `providers`, including NLP parsers, concept identification modules, synonym expansion modules, classifiers, etc.

Resources. The Yang et al. (2015) system uses LingPipe and ClearNLP[7] (Choi and Palmer, 2011) to parse the questions and relevant snippets using models applicable to generic English texts as well as biomedical texts, e.g. the parser models trained on the CRAFT treebank (Verspoor et al., 2012). It uses the named entity recognition (NER) module from LingPipe[8] trained on the GENIA corpus (Kim et al., 2003) and MetaMap annotation component (Aronson, 2001) to identify the biomedical concepts, and further uses UMLS Terminology Services (UTS)[9] to identify concepts and retrieve synonyms. It uses the official GoPubMed services for concept retrieval, and a local Lucene[10] index for document retrieval and snippet retrieval. LibLinear[11] (Fan et al., 2008) is used to train classifiers to predict answer types to the questions and estimate the relevance scores of candidate answers.

Answer Type Prediction. To identify the gold standard labels for the existing Q/A pairs used for training, the Yang et al. (2015) system employs UTS to retrieve the semantic types for each gold standard exact answer. A number of linguistic and semantic features are extracted from the tokens and concepts, including the lemma form of each token, the semantic type of each concept in the question, the dependency label of each token, combination of semantic type labels and dependency labels, etc., where the concepts are identified from MetaMap, LingPipe NER, and Apache OpenNLP Chunker[12] (noun phrases). A multiclass Logistic Regression classifier is trained using the LibLinear tool (Fan et al., 2008).

Candidate Answer Generation. Depending on the question type (general factoid/list question, `CHOICE` question, or `QUANTITY` question), the Yang et al. (2015) system applies different strategies to generate candidate answers. For general factoid/list questions, it generates a candidate answer using each concept identified by one of three

[4] https://github.com/oaqa/cse-framework/

[5] http://yaml.org/

[6] https://github.com/oaqa/baseqa/

[7] http://www.clearnlp.com

[8] http://alias-i.com/lingpipe/

[9] https://uts.nlm.nih.gov/home.html

[10] https://lucene.apache.org/

[11] http://www.csie.ntu.edu.tw/~cjlin/liblinear/

[12] https://opennlp.apache.org/

concept identification approaches. For `CHOICE` questions, it first identifies the "or" token in the question and its head token, which is most likely the *first option* in the list of candidate answers, and then finds all the children of the first option token in the parse tree that have a dependency relation of `conj`, which are considered to be *alternative options*. For `QUANTITY` questions, it identifies all the tokens that have a POS tag of `CD` in all relevant snippets.

Candidate Answer Scoring and Pruning. The Yang et al. (2015) system extends the approach used by Weissenborn et al. (Weissenborn et al., 2013) and defines 11 groups of features to capture how likely each candidate answer is the true answer for the question from different aspects, which includes answer type coercion, candidate answer occurrence count, name count, average overlapping token count, stopword count, overlapping concept count, token and concept proximity, etc. A Logistic Regression classifier is used to learn the scoring function, where the class is weighted by their frequencies. A simple threshold based pruning method is trained from the development dataset and applied to the list questions.

Besides incorporating a larger development data set, our OAQA system extends the Yang et al. (2015) system by integrating additional biomedical and general-purpose NLP annotators, and introducing trainable modules in more stages of the pipeline, such as using supervised methods in search result reranking, answer reranking, and yes/no answer prediction, which we will detail in the following sections. The architecture diagrams are illustrated in Figures 1, 2, and 3 in Appendix.

3 Concept Identification

We use the MetaMap and LingPipe concept identification modules with the GENIA model from Yang et al. (2015). However, due to the excessive noise introduced from the Apache OpenNLP Chunker based method, which extracts all noun phrases, we discard this approach. In addition, we integrate the TmTool biomedical concept identification RESTful service (Wei et al., 2016) for both the semantic type labeling of gold standard exact answers and question/snippet annotation, and also use C-Value (Frantzi et al., 2000), a frequent phrase mining method, to extract potential out-of-vocabulary multi-word terms.

3.1 TmTool for Annotating Questions, Snippets, and Answer Texts

The TmTool provides a standard Web service interface to annotate biomedical concepts using a number of state-of-the-art biomedical NLP parsers, which includes GNormPlus/SR4GN (for genes and species), tmChem (for chemicals), DNorm (for diseases), and tmVar (for mutations). Although it can only identify biomedical concepts belonging to any of these categories, they account for a great portion of the concepts used in the BioASQ corpus. In addition, many of these parsers utilize morphological features to estimate the likelihood of a term being a biomedical concept, rather than relying on an existing ontology like MetaMap, which makes it complementary to the existing tools in Yang et al. (2015).

TmTool supports three data exchange formats: PubTator (tab-separated), BioC (XML) and PubAnnotation (JSON). Since the PubTator format does not support DNorm annotation, and BioC format sometimes causes a single-sentence request to timeout (no response after 20 minutes) , we chose the robustest PubAnnotation format. We also found that the offsets returned from the TmTool RESTful service might not align well with original request texts, especially with tmChem trigger, and hence we implement an `escape` method to convert the texts into a TmTool compatible format by replacing some non-ASCII characters with their normalized forms, and removing special characters.

We use the TmTool to identify the biomedical concepts and annotate their semantic types from both the questions (in Phases A and B) and the relevant snippets (in Phase B) in the same manner as MetaMap. As the semantic type set of concepts has expanded to include TmTool concept types, the answer type prediction module should also be able to predict these additional semantic types. Therefore, we also use the TmTool to label the semantic types of the gold standard exact answers. In particular, we concatenate all the exact answers of each factoid or list question using commas, and send the concatenated string to the TmTool service, instead of each exact answer at a time. For example, if the gold standard exact answer is a list of strings: "NBEAL2", "GFI1B", "GATA1", then a single string "NBEAL2, GFI1B, GATA1" will be sent.

3.2 C-Value for Extracting Multi-Word Terms from Snippets

We treat the relevant snippets provided for each question in Phase B as a corpus, and we hypothesize that if a multi-word phrase is frequent in the corpus, then it is likely a meaningful concept. In order to extract not only high-frequency terms but also high-quality terms, a C-Value criterion (Frantzi et al., 2000) is introduced, which subtracts the frequency of its super terms from a term's own frequency. In this way, it returns the longer multi-word terms if two candidate terms overlap and have the same frequency. This approach only applies to a corpus, rather than a single sentence. Therefore, we only use this method to extract concepts from snippets. In the future, we may consider to collect a corpus relevant to the question, in order to apply the same idea to questions.

4 Collective Answer Reranking

We employ a collective answer reranking method aiming to boost the low-ranked candidate answers which share the same semantic type with high-ranked candidate answers for list questions, and use an adaptive threshold for pruning. The intuition is that list questions always ask for a list of concepts that have the same properties, which implies that the concepts usually have the same semantic types (e.g. all of them should be gene names). After the answer scoring step where a confidence score is assigned to each candidate answer *individually*, we can imagine the top candidate answers might have mixed types. For example, in a situation where the second answer is a disease, but the rest of the top-5 answers are all gene names, we should expect that the second answer should be down-ranked.

We use the same labels used for training the candidate answer scoring model, but incorporate features that measure how similar each answer is to the other top-ranked answers, which are detailed in Table 1. The *token distance* counts the number of intermediate tokens between the candidate answer tokens in the snippet text, and *Levenshtein edit distance* and *shape edit distance* measure the morphological similarities between the answer texts. *Common semantic type count* should "promote" the candidate answers that have a large number of semantic types in common with the top ranked answers.

For each candidate answer, we calculate a feature value, according to Table 1, for each other candidate answer in the input candidate list, and then we calculate the max/min/avg value corresponding to the top-k candidate answers. We use 1, 3, 5, 10 for k, and use Logistic Regression to train a binary classifier by down-sampling the negative instances to balance the training set. In addition to list questions, we also apply the method to factoid questions. In Section 7, we observe if the hypothesis also holds for factoid questions.

5 Learning to Answer Yes/No Questions

We consider the yes/no question answering problem as a binary classification problem, which allows to prioritize, weight, and blend multiple pieces of evidence from various approaches using a supervised framework. We list the sources of evidence (features) integrated into the system.

"Contradictory" concept. First, we hypothesize that if a statement is wrong, then the relevant snippets should contain some statements that are contradictory to the original statement, with some mentions of "contradictory" concepts or "antonyms". To identify pairs of contradictory concepts or antonyms is difficult given the resources that we have. Instead, we try to identify all the different concepts in the snippets that have the same semantic type as each concept in the original statement. For a given concept type, the more the unique concepts are found in both question and relevant snippets, or the less the concepts in the questions are found in the snippets, the more likely the original statement is wrong.

Formally, for a concept type t, we calculate a "contradictory" score as follows:

$$\frac{\sum_{s \in S} \sum_{c \in s} [\text{type}(c) = t]}{\sum_{c \in q} [\text{type}(c) = t] + \sum_{s \in S} \sum_{c \in s} [\text{type}(c) = t]}$$

where S is the set of snippets, q is the question, c is a concept mention, and $[\text{type}(c) = t]$ takes 1 if the concept c is type t and 0 otherwise. We derive the aggregated contradictory score from the concept type level scores using max/min/average statistics. We calculate a number of similar statistics to estimate how likely each snippet contradicts the original statement.

Overlapping token count. In case the concept identification modules fail to identify important concepts in either the original questions or relevant snippets, we also consider the difference of

No.	Feature
1	the original score from the answer scoring prediction
2	min/max/avg token distance between each pair of candidate answer occurrences
3	min/max/avg Levenshtein edit distance between each pair of candidate answer variant names
4	min/max/avg number (and percentage) of semantic types that each pair of candidate answers have in common
5	min/max/avg edit distance between each pair of candidate answer variant names after transformed into their *shape* forms (i.e. upper-case letters are replaced with 'A', lower-case letters are replaced with 'a', digits are replaced with '0', and all other characters are replaced with '-'.)

Table 1: Collective Answer Reranking Features

No.	Feature
1	"contradictory" concept count in the relevant snippets
2	overlapping token count in the relevant snippets
3	expected answer count in the relevant snippets
4	sentiment analysis via positive and negative word count of each relevant snippet
5	negation word count of each relevant snippet
6	question inversion

Table 2: Yes/No Question Answering Features

token mentions between the original question and the relevant snippets, instead of concepts.

Expected answer count. Not all concepts and tokens are equally important in the original questions. We find that many times the focus of a yes/no question is the last concept mention, which we denote as the *expected answer*. We count the frequency (and the percentage) that the expected answer is mentioned in the relevant snippets, as well as the frequency that concepts of the same type are mentioned.

Positive / negative / negation word count. Sometimes, an explicit sentiment is expressed in the relevant snippets to indicate how confident the author believes a statement is true or false. We use a simple dictionary [13] based method (Hu and Liu, 2004) for sentiment analysis, and we count whether and how many times each positive / negative word is mentioned in each snippet, then aggregate across the snippets using min / max / average. We also use a list of common English negation words[14] for negation detection, for simplicity. Intuitively, a high overlapping count with a high negative or negation count indicates that the original statement tends to be incorrect.

Question inversion. The question inversion method (Kanayama et al., 2012) answers a yes/no question by first converting it to a factoid question, then applies an existing factoid question answering pipeline to generate a list of alternate candidate answers, and finally evidence each candidate answer and rank them. If the expected answer in the original question is also ranked at the top among all candidates for the factoid question, then the statement is true.

In our system, we first assume the last concept mention corresponds to the expected answer. Therefore, its concept type(s) are also the answer type(s) of the factoid question, and all the synonyms of the concept are the answer variants. After the token(s) and concept mention(s) covered by the expected answer are removed from the original question and the question type is changed to `FACTOID`, we use the candidate answer generation and scoring pipeline for the factoid QA to generate and rank a list of candidate answers. Since annotating additional texts is computationally expensive, we do not retrieve any relevant snippets for the converted factoid questions, instead we only use the relevant snippets of the original yes/no questions (provided as part of the Phase B input). The rank and the score of the expected answer are used as question inversion features for yes/no question training.

We use a number of classifiers, e.g. Logistic Regression, Classification via Regression (Frank et

[13] `http://www.cs.uic.edu/~liub/FBS/opinion-lexicon-English.rar`

[14] `http://www.enchantedlearning.com/wordlist/negativewords.shtml`

al., 1998), Simple Logistic (Landwehr et al., 2005) using LibLinear and Weka[15] tools (Hall et al., 2009), after we down-sampled the positive ("yes") instances. In Section 7, we report not only the performance of each method in terms of accuracy, but also accuracy on the "yes" questions and the "no" questions, since on an imbalanced dataset, a simple "all-yes" method is also a "strong" baseline.

6 Retrieval Result Reranking via Relevance Classification

For relevant document, concept, and snippet retrieval, we first retrieve a list of 100 candidate results, then we define a set of features to estimate the relevance of each candidate result and employ a standardized interface to incorporate these features to rerank the retrieval result, which is different from Yang et al. (2015), where each stage employs a different retrieval / reranking strategy.

First, we replace the GoPubMed services with local Lucene indexes as the response time is estimated to be at least 20 times faster, although the task performance could be slightly worse (.2762 in terms of MAP using the GoPubMed concept Web service vs. .2502 using the local Lucene index in our preliminary experiment for concept retrieval). The concept Lucene index was created by fusion of the same biomedical ontologies used by the GoPubMed services, where we create 3 text fields: concept name, synonyms, and definition, and 2 string fields: source (Gene Ontology, Disease Ontology, etc) and URI. The document Lucene index was created from the latest MEDLINE Baseline corpus[16] using Lucene's StandardAnalyzer. After a list of documents are retrieved and segmented into sections and sentences, the snippet Lucene index is then built in memory on-the-fly at the sentence level. The search query is constructed by concatenating all synonyms of identified concepts (enclosed in quotes) and all tokens that are neither covered by any concept mentions nor are stop words, where the most 5,000 common English words[17] are used as the stop list. Then, the query searches all text fields.

The standardized search result reranking interface allows each retrieval task to specify different scoring functions (features). The features that we used for concept, document, and snippet retrieval are listed in Table 3. For example, during concept search result reranking, we can check if each candidate concept is also identified in the question by a biomedical NER. During snippet reranking, we can also incorporate the meta information, such as section label (title, abstract, body text, etc.), offsets in the section, and length of each snippet. In the candidate retrieval step, we have used a query that combines all non stop words and concepts identified by all biomedical concept annotators, in order to guarantee high recall. However, it does not optimize the precision. For example, some annotators/synonym expansion services may falsely identify concepts and introduce noisy search terms, and some search fields tend to be less informative than others. Therefore, in the reranking step, we employ various query formulation strategies, e.g. only within certain text fields and/or only using a subset of concept annotators, and consider the search score and rank of each candidate search result as features.

For this year's evaluation, we use Logistic Regression to learn relevance classifiers for all the reranking tasks, after negative instances are down-sampled to balance the training set. In the future, we can also integrate learning-to-rank modules.

7 Results

Besides the proposed methods described in Sections 3 to 6, we also made a few minor changes to the Yang et al. (2015) system, including

1. separating the texts in the parentheses in all gold standard exact answers as synonyms before gold standard semantic type labeling and answer type prediction training,
2. introducing the "null" type for the exact answer texts if neither of the two concept search providers (TmTool or UTS) can identify,
3. and adding nominal features (e.g. answer type name, concept type name, etc.) in addition to existing numeric features (e.g. count, distance, ratio, etc.) for candidate answer scoring.

In this section, we first report the evaluation results using the held-out BioASQ 3B Batch 5 subset, and then we conduct a manual analysis using BioASQ 4B dataset for our yes/no question answering method.

We first compare the retrieval results (Phase A) In Table 4. We can see that the proposed retrieval

[15] `http://www.cs.waikato.ac.nz/ml/weka/`
[16] `https://mbr.nlm.nih.gov/`
[17] `http://www.wordfrequency.info/`

No.	Feature
Concept	
1	the original score from the concept retrieval step
2	overlapping concept count between the retrieval results and mentions annotated in the question
3	retrieval scores using various query formulation strategies
Document	
1	the original score from the document retrieval step
2	retrieval scores using various query formulation strategies
Snippet	
1	the score of the containing document
2	meta information, including the section label (abstract or title), binned begin/end offsets, binned length of the snippet, etc.
3	retrieval scores using various query formulation strategies

Table 3: Retrieval Result Reranking via Relevant Classification Features

Method	MAP	F1	Precision	Recall
Concept				
LR	**.3216**	.0297	.0154	.5504
NO	.2361			
Document				
LR	**.1364**	.0462	.0284	.2709
NO	.1003			
Snippet				
NO	**.1073**	.0147	.0079	.3015
LR	.0826			

Table 4: Evaluation results on BioASQ 3B Batch 5 Phase A subset. LR represents a Logistic Regression based reranking method is used, and NO means no operation is performed, i.e. original retrieval scores are used.

result reranking method via Logistic Regression improves the performance of concept and document retrieval, but not snippet retrieval, which may be due to the fact that the input candidate snippets have been reranked using a similar set of features at the document reranking step, and no further information is provided during the subsequent snippet reranking step.

The exact answer generation results (Phase B) are shown in Table 5. We see that the best configuration for factoid question answering in terms of MRR is keeping the original feature set with no collective reranking. However, if additional features are used, then the collective reranking method can improve the performance, and achieve the highest lenient accuracy score.

To answer list questions, we tune the thresholds (hard threshold, or TP and ratio threshold, or RP) and report the results from the thresholds that maximize the F1 score. Although the best F1 score is achieved by incorporating additional features without collective reranking, and using a ratio-based pruning method, all other configurations without collective reranking have the lowest performance. In addition, we can see that additional features improve the performance in general, and after carefully tuning of the threshold and the ratio in the pruning step, we can achieve the same level of performance. We hypothesize that the proposed method (CR + RP) can renormalize the answer scores and is thus more robust than the baseline system (NO + TP) in the sense that the performance of the former approach is less sensitive to the predefined threshold, although the latter can sometimes outperform the former when the threshold is carefully tuned. We submitted two runs in BioASQ 4B Batch 5 evaluation: `oaqa-3b-5` and `oaqa-3b-5-e` for the proposed and baseline methods respectively (using the same thresholds), and initial evaluation result confirms our hypothesis.

Due to the imbalance between "yes" and "'no" questions, we report the mean negative and positive accuracy scores in addition to the overall accuracy for yes/no question answering. We can see the performance is very sensitive to the choice of the classifier. Using the same set of features, ClassificationViaRegression achieves the highest performance, with both negative and positive accuracy scores greatly above 0.5 (random). All other methods tend to predict "yes", which results in a high positive accuracy but a low (below 0.5) negative accuracy score.

Factoid			
Method	**Len.Ac.**	**MRR**	**Str.Ac.**
OF + NO	.5000	**.3843**	.3182
OF + CR	.4545	.3791	.3182
AF + CR	**.5455**	.3732	.2727
AF + NO	.5000	.3689	.2727
List			
Method	**F1**	**Precision**	**Recall**
AF + NO + RP	**.4291**	**.4449**	.4593
AF + CR + RP	.4246	.4045	**.4864**
AF + CR + TP	.3969	.4100	.4267
OF + CR + TP	.3704	.4231	.3645
OF + CR + RP	.3629	.3654	.3874
AF + NO + TP	.3463	.3840	.3677
OF + NO + RP	.3460	.3188	.4431
OF + NO + TP	.1461	.2639	.1183
Yes/No			
Method	**Ac.**	**Neg.Ac.**	**Pos.Ac.**
CVR	**.7143**	**.7778**	.6842
SL	**.7143**	.4444	**.8421**
AY	.6786	.0000	1.0000
LR	.5357	.2222	.6842

Table 5: Evaluation results on BioASQ 3B Batch 5 Phase B subset. OF and AF represent Original or Additional features are used in training and predicting answer scorers for factoid and list questions. CR represents the Logistic Regression based Collective Reranking is used. TP means a hard Threshold is used to prune the answer, whereas RP uses the relative Ratio to the maximum score. CVR and SL are ClassificationViaRegression and SimpleLogistic classifiers from the Weka toolkit. AY means "All-Yes", a simple but strong baseline, in terms of accuracy.

8 Analysis

From Section 7, we see that, despite integration of various sources of evidence, the current yes/no question answering system is still unreliable. We conducted a manual analysis of our yes/no question answering method using BioASQ 4B dataset based on our own judgment of yes or no, which may not be consistent with the gold standard.

We found the BioASQ 4B dataset is more imbalanced than the development dataset, where we only identified six questions from all five test batches that have a "no" answer. We applied the proposed yes/no QA method to the six questions. Among these questions, three are correctly predicted (namely, "Is macitentan an ET agonist?", "Does MVIIA and MVIIC bind to the same calcium channel?", and "Is the abnormal dosage of ultraconserved elements disfavored in cancer cells?"), and the answers to the other three questions are wrong. We conduct an error analysis for the false positive predictions.

The first false positive question is "Are adenylyl cyclases always transmembrane proteins?" The key to this question is the recognition of the contradictory concept pair "transmembrane" and "soluble" or "transmembrane adenylyl cyclase (tmAC)" and "soluble AC". This requires first correctly identifying both terms as biomedical concepts and then assigning correct semantic type labels to them, where the latter can only be achieved using MetaMap and TmTool. MetaMap correctly identified "transmembrane proteins" in the question and assigned a semantic label of "Amino Acid, Peptide, or Protein", and identified "soluble adenylyl cyclase" in the snippet and assigned a semantic label of "Gene or Genome". Due to the mismatch of semantic types "Amino Acid, Peptide, or Protein" and "Gene or Genome", the system fails to recognize the contradiction.

In fact, we found that the same problem also happened during the answer type prediction and answer scoring steps, e.g. the question may be predicted to ask for a "Gene or Genome", but the candidate answer is often labeled as "Amino Acid, Peptide, or Protein" by MetaMap/UTS. Because of the interchangeable use of "Amino Acid, Peptide, or Protein" and "Gene or Genome" terms, we might consider to treat them as one type. Moreover, the universal quantifier "always" also plays an important role, in contrast to a question with an existential quantifier such as "sometimes", which the current system has not captured yet. However, this is not the main reason of the failure, since we assume the relevant snippets will rarely mention "soluble AC" if the question asks for whether "transmembrane" exists.

The second false positive question is "Can chronological age be predicted by measuring telomere length?" This should be an easy one, because we can find a negation cue "cannot" in the snippet "telomere length measurement by real-time quantitative PCR cannot be used to predict age of a person". The system integrates two types of negation cue related features: the negation cue count and the existence of a particular negation cue. We found the system correctly identified and counted the negation cue. Therefore, we suspect the classifier did not optimize the combination of features. Furthermore, we need to observe whether our hypothesis that the gold standard answer (yes or no) is strongly correlated with the negation word occurrence in the relevant snippets is true using the development set.

The third false positive question is "Does the 3D structure of the genome remain stable during cell differentiation?" The key to this question is the word "stable", which requires biomedical, esp. genomics, knowledge to understand what "stable" means in the context of genome structure. The word "stable" is mentioned in one of the snippets "the domains are stable across different cell types", which however does not answer the question. Useful contradictory keywords that we find in the relevant snippets include "reorganization", "alteration", "remodelling", etc. MetaMap/UTS identified "stable" as a concept of semantic type "Qualitative Concept", whereas it labeled "reorganization" as a "Idea or Concept" and missed "alternation" and "remodelling". It suggests that our contradictory concept based method works the best if the focus is factoid (entities), but the current knowledge base can hardly support identification of contradictory properties or behaviors.

We focus on 4B Batch 5 subset for error analysis of false negative examples. In fact, the cases for false negative questions are more diverse, which makes it more difficult to find the causes of failures. One reason is that some snippets contain multiple sentences or clauses, and only one is crucial to answer the question, while others can negatively influence the results. For example, the snippet "OATP1B1 and OATP1B3-mediated transport of bilirubin was confirmed and inhibition was determined for atazanavir, rifampicin, indinavir, amprenavir, cyclosporine, rifamycin SV and saquinavir." has two clauses, but the second one ("and inhibition..."), although is not relevant to the question, introduces other chemical names that confuse the classifier. Another problem is lack of understanding of specificity and generality between concepts, e.g. "encephalopathy" in the question is considered a different concept from "Wernicke encephalopathy" mentioned in the snippets, both belonging to the same disease category. The classifier believed another disease name is mentioned to contradict the statement.

We found that yes/no questions are more difficult to answer than factoid and list questions, since there can be many different ways to support or oppose a statement. Although the problem can be simply viewed as a binary classification problem, due to the fact that a limited number of relevant snippets are provided, simple token or phrase level retrieval and statistics can hardly solve the problem. Instead, we believe that reliably answering yes/no questions requires deeper linguistic and semantic understanding of the questions and relevant snippets, which includes leveraging semantic networks of concepts to identify antonyms, hypernyms, and hyponyms, and utilizing dependency relations between the concepts, as well as sentiment analysis of the facts.

9 Conclusion

This paper describes the OAQA system evaluated in the BioASQ 4B Question Answering track. We first present the overall architecture of the system, and then focus on describing the main differences from the Yang et al. (2015) system, including two concept identification modules: TmTool and C-value based multi-word term extractor, collective answer reranking, yes/no question answering approach, and a standardized retrieval result reranking method via relevant classification. We report our initial evaluation results on 3B Batch 5 subset show the effectiveness of the proposed new methods, and since the yes/no question answering approach is unsatisfactory, we further conduct an error analysis for yes/no questions using 4B subset.

As we mention in earlier sections, to further improve the retrieval performance, we may use learning-to-rank methods to rerank the retrieval results. For exact answer generation, esp. for yes/no questions, we believe a deeper linguistic and semantic analysis of both questions and relevant snippets are necessary. Our preliminary experiment suggested that the word2vec (Mikolov et al., 2013) based method did worse than the KB based method in modeling the semantics of entities. We plan to study whether the former is complementary to the latter in representing the semantics of biomedical properties and event mentions.

Acknowledgments

We thank Ying Li, Xing Yang, Venus So, James Cai and the other team members at Roche Innovation Center New York for their support of OAQA and biomedical question answering research and development.

References

Alan R Aronson. 2001. Effective mapping of biomedical text to the umls metathesaurus: the metamap program. In *Proceedings of the AMIA Symposium*, page 17. American Medical Informatics Association.

Jinho D Choi and Martha Palmer. 2011. Getting the most out of transition-based dependency parsing. In *Proceedings of the 49th Annual Meeting of the Association for Computational Linguistics: Human Language Technologies: short papers-Volume 2*, pages 687–692. Association for Computational Linguistics.

Rong-En Fan, Kai-Wei Chang, Cho-Jui Hsieh, Xiang-Rui Wang, and Chih-Jen Lin. 2008. Liblinear: A library for large linear classification. *J. Mach. Learn. Res.*, 9(Aug):1871–1874.

David Ferrucci, Eric Nyberg, James Allan, Ken Barker, Eric Brown, Jennifer Chu-Carroll, Arthur Ciccolo, Pablo Duboue, James Fan, David Gondek, et al. 2009. Towards the open advancement of question answering systems. Technical Report RC24789 (W0904-093), IBM Research Division.

Eibe Frank, Yong Wang, Stuart Inglis, Geoffrey Holmes, and Ian H Witten. 1998. Using model trees for classification. *Machine Learning*, 32(1):63–76.

Katerina Frantzi, Sophia Ananiadou, and Hideki Mima. 2000. Automatic recognition of multi-word terms:. the c-value/nc-value method. *International Journal on Digital Libraries*, 3(2):115–130.

Elmer Garduno, Zi Yang, Avner Maiberg, Collin McCormack, Yan Fang, and Eric Nyberg. 2013. Cse framework: A uima-based distributed system for configuration space exploration. In *UIMA@GSCL'2013*, pages 14–17.

Mark Hall, Eibe Frank, Geoffrey Holmes, Bernhard Pfahringer, Peter Reutemann, and Ian H Witten. 2009. The weka data mining software: an update. *ACM SIGKDD explorations newsletter*, 11(1):10–18.

Minqing Hu and Bing Liu. 2004. Mining and summarizing customer reviews. In *Proceedings of the tenth ACM SIGKDD international conference on Knowledge discovery and data mining*, pages 168–177. ACM.

Hiroshi Kanayama, Yusuke Miyao, and John Prager. 2012. Answering yes/no questions via question inversion. In *COLING*, pages 1377–1392. Citeseer.

J-D Kim, Tomoko Ohta, Yuka Tateisi, and Junichi Tsujii. 2003. Genia corpusa semantically annotated corpus for bio-textmining. *Bioinformatics*, 19(suppl 1):i180–i182.

Niels Landwehr, Mark Hall, and Eibe Frank. 2005. Logistic model trees. *Machine Learning*, 59(1-2):161–205.

Tomas Mikolov, Kai Chen, Greg Corrado, and Jeffrey Dean. 2013. Efficient estimation of word representations in vector space. *arXiv preprint arXiv:1301.3781*.

George Tsatsaronis, Georgios Balikas, Prodromos Malakasiotis, Ioannis Partalas, Matthias Zschunke, Michael R Alvers, Dirk Weissenborn, Anastasia Krithara, Sergios Petridis, Dimitris Polychronopoulos, et al. 2015. An overview of the bioasq large-scale biomedical semantic indexing and question answering competition. *BMC Bioinformatics*, 16(1):138.

Karin Verspoor, Kevin Bretonnel Cohen, Arrick Lanfranchi, Colin Warner, Helen L Johnson, Christophe Roeder, Jinho D Choi, Christopher Funk, Yuriy Malenkiy, Miriam Eckert, et al. 2012. A corpus of full-text journal articles is a robust evaluation tool for revealing differences in performance of biomedical natural language processing tools. *BMC bioinformatics*, 13(1):1.

Chih-Hsuan Wei, Robert Leaman, and Zhiyong Lu. 2016. Beyond accuracy: creating interoperable and scalable text-mining web services. *Bioinformatics*, pages 1–4.

Dirk Weissenborn, George Tsatsaronis, and Michael Schroeder. 2013. Answering factoid questions in the biomedical domain. In *BioASQ'2013*.

Zi Yang, Elmer Garduno, Yan Fang, Avner Maiberg, Collin McCormack, and Eric Nyberg. 2013. Building optimal information systems automatically: Configuration space exploration for biomedical information systems. In *CIKM'2013*, pages 1421–1430.

Zi Yang, Niloy Gupta, Xiangyu Sun, Di Xu, Chi Zhang, and Eric Nyberg. 2015. Learning to answer biomedical factoid & list questions: Oaqa at bioasq 3b. In *CLEF'2015 (Working Note)*.

Appendix

Listing 1: ECD main descriptor for factoid and list QA in batch 5 in Phase B

```
# execute
#       mvn exec:exec -Dconfig=bioasq.phase-b-test-factoid-list
# to test the pipeline

configuration:
  name: phase-b-test-factoid-list
  author: ziy

persistence-provider:
  inherit: baseqa.persistence.local-sqlite-persistence-provider

collection-reader:
  inherit: baseqa.collection.json.json-collection-reader
  dataset: BIOASQ-QA
  file:
    - input/4b-5-b.json
  type: [factoid, list]
  persistence-provider: |
    inherit: baseqa.persistence.local-sqlite-persistence-provider

pipeline:
  - inherit: ecd.phase
    name: question-parse
    options: |
      - inherit: bioqa.question.parse.clearnlp-bioinformatics

  - inherit: ecd.phase
    name: question-concept-metamap
    options: |
      - inherit: bioqa.question.concept.metamap-cached

  - inherit: ecd.phase
    name: question-concept-tmtool
    options: |
      - inherit: bioqa.question.concept.tmtool-cached

  - inherit: ecd.phase
    name: question-concept-lingpipe-genia
    options: |
      - inherit: bioqa.question.concept.lingpipe-genia

  - inherit: ecd.phase
    name: question-focus
    options: |
      - inherit: baseqa.question.focus

  - inherit: ecd.phase
    name: passage-to-view
    options: |
      - inherit: baseqa.evidence.passage-to-view

  - inherit: ecd.phase
    name: evidence-parse
    options: |
      - inherit: bioqa.evidence.parse.clearnlp-bioinformatics

  - inherit: ecd.phase
    name: evidence-concept-metamap
    options: |
      - inherit: bioqa.evidence.concept.metamap-cached

  - inherit: ecd.phase
    name: evidence-concept-tmtool
    options: |
      - inherit: bioqa.evidence.concept.tmtool-cached

  - inherit: ecd.phase
    name: evidence-concept-lingpipe-genia
    options: |
      - inherit: bioqa.evidence.concept.lingpipe-genia

  - inherit: ecd.phase
    name: evidence-concept-frequent-phrase
    options: |
      - inherit: baseqa.evidence.concept.frequent-phrase

  - inherit: ecd.phase
    name: concept-search-uts
    options: |
      - inherit: bioqa.evidence.concept.search-uts-cached

  - inherit: ecd.phase
    name: concept-merge
    options: |
      - inherit: baseqa.evidence.concept.merge

  - inherit: ecd.phase
    name: answer-type
    options: |
      - inherit: bioqa.answer_type.predict-liblinear-null

  - inherit: ecd.phase
    name: answer-generate
    options: |
      - inherit: bioqa.answer.generate.generate

  - inherit: ecd.phase
    name: answer-modify
    options: |
      - inherit: baseqa.answer.modify.modify

  - inherit: ecd.phase
    name: answer-score
    options: |
      - inherit: bioqa.answer.score.predict-liblinear

  - inherit: ecd.phase
    name: answer-collective-score
    options: |
      - inherit: bioqa.answer.collective_score.predict-liblinear
      - inherit: base.noop

  - inherit: ecd.phase
    name: answer-prune
    options: |
      - inherit: baseqa.answer.modify.pruner

#  - inherit: baseqa.cas-serialize

post-process:
  # submission
  - inherit: bioasq.collection.json.json-cas-consumer
```

Listing 2: ECD main descriptor for yes/no QA in batch 5 in Phase B

```
# execute
#       mvn exec:exec -Dconfig=bioasq.phase-b-test-yesno
# to test the pipeline

configuration:
  name: phase-b-test-yesno
  author: ziy

persistence-provider:
  inherit: baseqa.persistence.local-sqlite-persistence-provider

collection-reader:
  inherit: baseqa.collection.json.json-collection-reader
  dataset: BIOASQ-QA
  file:
    - input/4b-5-b.json
  type: [yesno]
  persistence-provider: |
```

```
    inherit: baseqa.persistence.
      local-sqlite-persistence-provider

pipeline:
  - inherit: ecd.phase
    name: question-parse
    options: |
      - inherit: bioqa.question.parse.
        clearnlp-bioinformatics

  - inherit: ecd.phase
    name: question-concept-metamap
    options: |
      - inherit: bioqa.question.concept.
        metamap-cached

  - inherit: ecd.phase
    name: question-concept-tmtool
    options: |
      - inherit: bioqa.question.concept.
        tmtool-cached

  - inherit: ecd.phase
    name: question-concept-lingpipe-genia
    options: |
      - inherit: bioqa.question.concept.
        lingpipe-genia

  - inherit: ecd.phase
    name: passage-to-view
    options: |
      - inherit: baseqa.evidence.passage-to-view

  - inherit: ecd.phase
    name: evidence-parse
    options: |
      - inherit: bioqa.evidence.parse.
        clearnlp-bioinformatics

  - inherit: ecd.phase
    name: evidence-concept-metamap
    options: |
      - inherit: bioqa.evidence.concept.
        metamap-cached

  - inherit: ecd.phase
    name: evidence-concept-tmtool
    options: |
      - inherit: bioqa.evidence.concept.
        tmtool-cached

  - inherit: ecd.phase
    name: evidence-concept-lingpipe-genia
    options: |
      - inherit: bioqa.evidence.concept.
        lingpipe-genia

  - inherit: ecd.phase
    name: evidence-concept-frequent-phrase
    options: |
      - inherit: baseqa.evidence.concept.
        frequent-phrase

  - inherit: ecd.phase
    name: concept-search-uts
    options: |
      - inherit: bioqa.evidence.concept.
        search-uts-cached

  - inherit: ecd.phase
    name: concept-merge
    options: |
      - inherit: baseqa.evidence.concept.merge

  - inherit: ecd.phase
    name: answer-yesno
    options: |
      - inherit: bioqa.answer.yesno.
        predict-weka-other
     - inherit: baseqa.answer.yesno.all-yes

post-process:
  # submission
  - inherit: bioasq.collection.json.
      json-cas-consumer
```

Listing 3: ECD main descriptor for retrieval in batch 5 in Phase A

```
# execute
#      mvn exec:exec -Dconfig=bioasq.phase-a-test
# to test the pipeline

configuration:
  name: phase-a-test
  author: ziy

persistence-provider:
  inherit: baseqa.persistence.
     local-sqlite-persistence-provider

collection-reader:
  inherit: baseqa.collection.json.
     json-collection-reader
  dataset: BIOASQ-QA
  file:
    - input/4b-5-a.json
  persistence-provider: |
    inherit: baseqa.persistence.
     local-sqlite-persistence-provider

pipeline:
  - inherit: ecd.phase
    name: question-parse
    options: |
      - inherit: bioqa.question.parse.
        clearnlp-bioinformatics

  - inherit: ecd.phase
    name: question-concept-metamap
    options: |
      - inherit: bioqa.question.concept.
        metamap-cached

  - inherit: ecd.phase
    name: question-concept-tmtool
    options: |
      - inherit: bioqa.question.concept.
        tmtool-cached

  - inherit: ecd.phase
    name: question-concept-lingpipe-genia
    options: |
      - inherit: bioqa.question.concept.
        lingpipe-genia

  - inherit: ecd.phase
    name: concept-search-uts
    options: |
      - inherit: bioqa.evidence.concept.
        search-uts-cached

  - inherit: ecd.phase
    name: concept-merge
    options: |
      - inherit: baseqa.evidence.concept.merge

  - inherit: ecd.phase
    name: abstract-query-primary
    options: |
      - inherit: baseqa.abstract_query.token-concept

  # concept
  - inherit: ecd.phase
    name: concept-retrieval
    options: |
      - inherit: bioqa.concept.retrieval.
        lucene-bioconcept

  - inherit: ecd.phase
    name: concept-rerank
    options: |
      - inherit: bioqa.concept.rerank.
        predict-liblinear

  # document
  - inherit: ecd.phase
    name: document-retrieval
    options: |
      - inherit: bioqa.document.retrieval.
        lucene-medline

  - inherit: ecd.phase
```

```
    name: document-rerank
    options: |
      - inherit: bioqa.document.rerank.predict-liblinear

  # snippet
  - inherit: ecd.phase
    name: passage-retrieval
    options: |
      - inherit: bioasq.passage.retrieval.document-to-passage

  - inherit: ecd.phase
    name: passage-rerank
    options: |
      - inherit: bioqa.passage.rerank.predict-liblinear
      - inherit: base.noop

post-process:
  # submission
  - inherit: bioasq.collection.json.json-cas-consumer
```

Listing 4: ECD component descriptor of `bioqa.answer.collective_score.liblinear-predict`

```
inherit: baseqa.learning_base.classifier-predict

candidate-provider: 'inherit: baseqa.answer.score.candidate-provider'
scorers: |
  - inherit: baseqa.answer.collective_score.scorers.original
  - inherit: baseqa.answer.collective_score.scorers.distance
  - inherit: baseqa.answer.collective_score.scorers.edit-distance
  - inherit: baseqa.answer.collective_score.scorers.type-coercion
  - inherit: baseqa.answer.collective_score.scorers.shape-distance
classifier: 'inherit: bioqa.answer.collective_score.liblinear-classifier'
feature-file: result/answer-collective-score-predict-liblinear.tsv
```

Listing 5: ECD component descriptor of `bioqa.answer.yesno.predict`

```
inherit: baseqa.answer.yesno.predict

scorers: |
  - inherit: baseqa.answer.yesno.scorers.concept-overlap
  - inherit: bioqa.answer.yesno.scorers.token-overlap
  - inherit: baseqa.answer.yesno.scorers.expected-answer-overlap
  - inherit: baseqa.answer.yesno.scorers.sentiment
  - inherit: baseqa.answer.yesno.scorers.negation
  - inherit: bioqa.answer.yesno.scorers.question-inversion
```

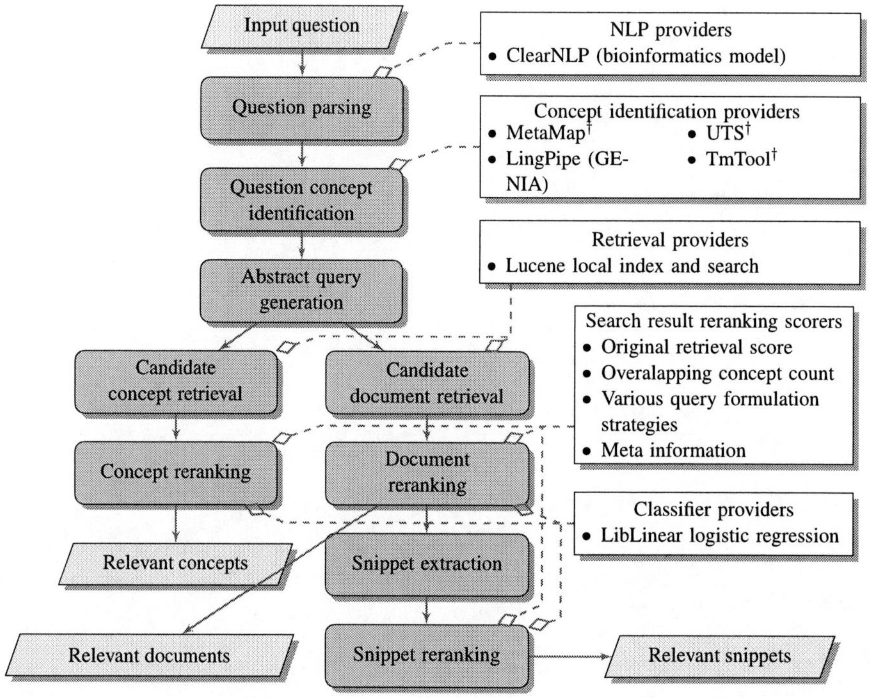

Figure 1: Retrieval (Phase A) pipeline diagram. † represents a provider that requires accessing external Web services.

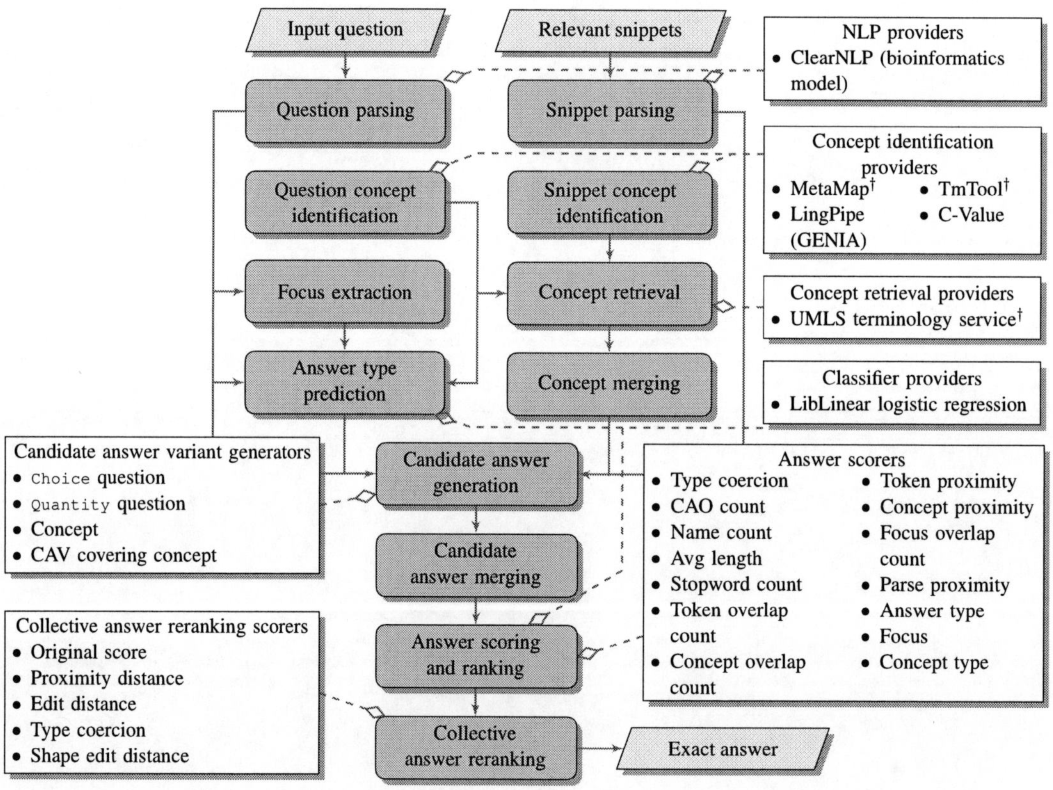

Figure 2: Factoid and list question answering (Phase B) pipeline diagram. † represents a provider that requires accessing external Web services.

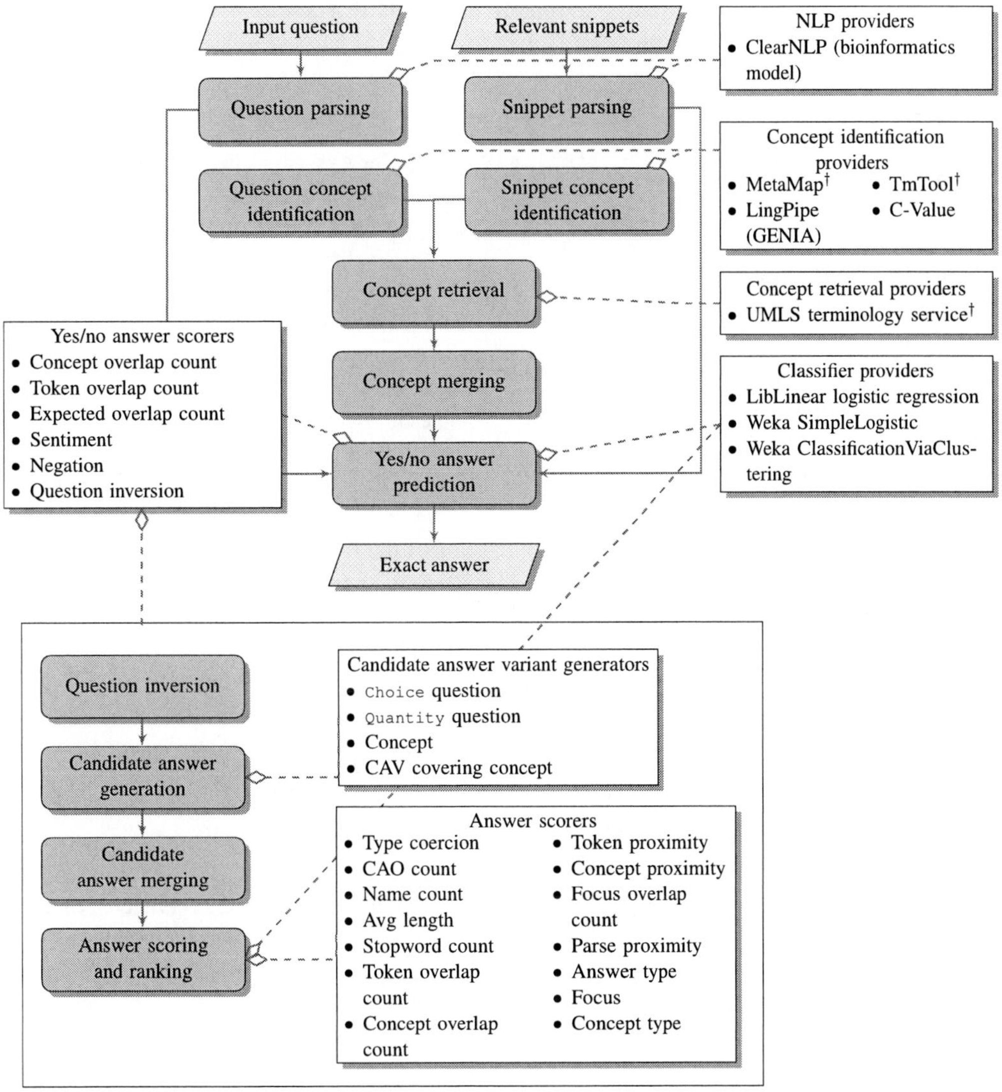

Figure 3: Yes/no question answering (Phase B) pipeline diagram. † represents a provider that requires accessing external Web services.

HPI Question Answering System in BioASQ 2016

Frederik Schulze, Ricarda Schüler, Tim Draeger, Daniel Dummer, Alexander Ernst, Pedro Flemming, Cindy Perscheid, Mariana Neves
Hasso-Plattner Institute
August-Bebel-Str. 88
Potsdam, Brandenburg, 14482 Germany
mariana.neves@hpi.de

Abstract

Question answering (QA) systems are crucial when searching for exact answers for natural language questions in the biomedical domain. Answers to many of such questions can be extracted from the 26 millions biomedical publications currently included in MEDLINE when relying on appropriate natural language processing (NLP) tools. In this work we describe our participation in the task 4b of the BioASQ challenge using two QA systems that we developed for biomedicine. Preliminary results show that our systems achieved first and second positions in the snippet retrieval sub-task and for the generation of ideal answers.

1 Introduction

The deluge of scientific publication in biomedicine requires tools for processing and searching precise information in real time. Question answering (QA) comes as an alternative to standard search engines system, e.g. PubMed[1], and provides precise and short answers for questions in natural language (Athenikos and Han, 2010; Neves and Leser, 2015). One of the advantages of QA systems is that the user does not need to be proficient in formulating queries in a way that the system can understand. Instead, a user may simply enter a question as they would pose it to another person and receive a answer in return. Thus, no formal training is required to use QA systems.

QA is one of the more complex applications of natural language processing (NLP) (Jurafsky and Martin, 2013). This is usually achieved through a three-steps architecture: (1) the users question must be processed so that a query can be generated; (2) this query is then used to find all relevant text passages from a large document collection; and (3) finally, the system generates the exact answer to the users question and/or a summary of the facts from these passages. Some QA systems already exist for the biomedical domain (Bauer and Berleant, 2012). However, none of them are capable of answering questions in real time, in part due to the large collections of documents involved in the task.

We describe our participation in the fourth edition of the BioASQ challenge[2] (Tsatsaronis et al., 2015), a community-based shared task which aims to evaluate the current solutions for a variety of QA sub-tasks. We submitted runs from two QA systems which were specifically developed for the biomedical domain. One of the system (HPI1) successfully participated in the previous editions of the BioASQ challenge (Neves, 2015) and our second system (HPI2) is described in this work. We relied on existing NLP functionality from a in-memory database (IMDB) and we extend it with new procedures tailored specifically to QA. We participated in the task 4b (Biomedical Semantic QA) which is split in two phases: (a) phase A: concept mapping and document, passage and RDF triples retrieval; and (b) phase B: exact and ideal (short summary) answers.

The next section presents a short description of our the HPI2 system, followed by the preliminary results that we obtained in the challenge and a short discussion about our performance and methods.

2 Data

We relied on two main resources when developing our QA system: the MEDLINE and the Unified

[1] http://www.ncbi.nlm.nih.gov/pubmed

[2] http://bioasq.org/

Proceedings of the 54th Annual Meeting of the Association for Computational Linguistics, pages 38–44,
Berlin, Germany, August 7-12, 2016.

Medical Language System (UMLS). In this section, we give a short overview on both resources.

2.1 MEDLINE

MEDLINE[3] is the main source for biomedical publications and grows continuously. We downloaded the publications from MEDLINE and integrated them into our local database. For the purposes of our QA system, an article consists of a title, an abstract and the main text. In this paper we refer only to titles and abstracts, as full papers are not considered in the current edition of the BioASQ challenge.

2.2 Unified Medical Language System

Extracting meaning out of biomedical documents is usually supported by manually curated dictionaries. These dictionaries contain words and phrases which are common to the biomedical domain. such dictionaries are used to map synonyms and abbreviations of terms to a common base term. Often, they also contain information to assign categories to terms. There are various terminologies for the biomedical domain, such as UMLS, SNOMED CT or MeSH.

UMLS[4] is a comprehensive database that combine various sources into a single knowledge base. It includes vocabularies mapping words and phrases onto a set of concepts. Each concept has an associated semantic type and group, which classifies the category of the concept, such as gene or disease.

In our QA system, UMLS was mainly used for named-entity recognition (NER), i.e., for extracting named-entities both in the question and in the document collection. Also in the context of NER, we used the UMLS semantic types to map the named-entities to their corresponding types. Finally, we also rely on UMLS to resolve synonyms, thus avoiding to miss important passages which include synonyms to the words in the questions. Abbreviations, in particular, are very frequent in biomedical documents.

3 Methods

Our QA is composed of many components (cf. Figure 1) which are included in three main steps, i.e., question processing, document retrieval, and answer processing. The later includes a two-step phase: exact answer extraction (not included in this paper) and summarization. Details for each component are described below.

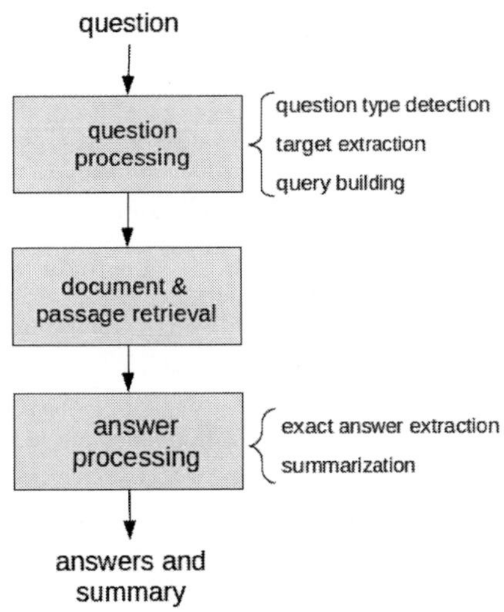

Figure 1: Work-flow of our question answering system.

3.1 Architecture

Our system was developed on top of a IMDB (SAP HANA database) (Plattner, 2013), which allows fast access of data directly from main memory, in contrast to processing data from files that reside on disk space, thus requiring loading data into main memory. The IMDB we used comes with built-in text analysis features, such as language detection, sentence splitting, tokenization, stemming, part-of-speech (POS) tagging, NER based on pre-compiled dictionaries, information extraction based on manually crafted rules, document indexing, approximate searching and sentiment analysis.

All textual resources (documents and questions) were added to the database and dictionaries of biomedical terms were created based on the UMLS terminology. Then we created the so-called full text index (FTI), i.e., an additional table which can be created for columns which contain text. Such an index can be created in many ways, we opted for two of them, namely: (a) a linguistic index, which contains all words from the original documents, as well as corresponding POS tags; and (b) a NER index, which contains all entities that were found based on the dictionary that was

[3] https://www.nlm.nih.gov/pubs/factsheets/medline.html

[4] https://www.nlm.nih.gov/research/umls/

previously built. In summary, from the linguistic FTI it is possible to retrieve information about sentence splitting, tokenization, stemming and POS tags, while the NER provides the named-entities.

3.2 Question processing

The first step in a question answering system is to analyze the input question. This step is composed of three components in our QA system: (a) question type detection, (b) target extraction, and (c) query building.

Question Type Detection. The question type can be either "yes/no", "factoid", "list" or "summary". It defines which kind of the answer the system needs to return. In this step, we split the question into words and and apply special rules to find the correct type, by considering question words and the structure (POS tags) of the question. Our approach is based on regular expressions, for instance, a questions beginning with an auxiliary verb is classified as yes/no-question. Although our QA system includes a component for detecting the question type, this step is not necessary in the BioASQ challenge because all question types are given.

Target Extraction. The second component of our question processing step extracts the target of the question, in case of factoid questions, and classifies it according to the UMLS semantic types[5], e.g, whether the question asks for a disease or a gene. This is an important information for the answer extraction step. We extract the headword using simple rules, for instance, the first noun after the question word (e.g., "what", "which"). For classifying the headwords according to the many UMLS semantic types, and inspired by (Huang et al., 2008), we relied on a machine learning (ML) approach based on the implementation of the Support Vector Machine (SVM) algorithm in the IMDB database. The features that we use were the headwords and the questions words. All headwords in the factoid questions were manually classified into the semantic types by one of the authors (MN) and this is the training data that was used in our experiments. During the process, several different features were evaluated, but they did not improved our results.

[5]`https://metamap.nlm.nih.gov/SemanticTypesAndGroups.shtml`

Query Building. Good query terms are important features when relying on a keyword-based search to find relevant documents for the question. For this purpose, we use all words, except for stopwords and question words (e.g., "what", "which").

3.3 Document and Passage Retrieval

The query that was built in the previous step was used in this step to find relevant documents and passages within the millions abstracts. We relied on the tf-idf method (Manning et al., 2008) as a basis and we adjusted it by various means to better fit the biomedical domain. We opted for the weighted tf-idf approach since our experiments showed that it provided up to 10% more recall than an equally weighted approach. We used a proximity measure to boosts a documents relevancy rating when it contains words from the query which appear close together. This measure searches for each possible word pair that appears in the query and applies a fixed rating increase for each such pair that is separated by a maximum of two words anywhere in the document.

We also consider the documents title in our approach. A documents titles relevancy was added to the documents relevancy in a weighted sum, thereby increasing the relevancy of documents with relevant titles. We also utilized a Jaccard-based word overlap measure between sentences in the document and in the question for the passage retrieval step. Our system first retrieves the 100.000 most relevant documents and then checks their sentences. This way we achieve a significant speed-up compared to calculating relevancy scores for all sentences in all documents. The document's total proximity score and the best sentence's word overlap score are then used to boost the initial tf-idf score. Their influence was tuned empirically on a test set of BioASQ questions and answers. Finally, our document and passage retrieval algorithms return a list of documents or passages, sorted by their relevance score.

3.4 Answer extraction

We only submitted ideal answers, i.e. short summaries, for the BioASQ challenge. Our approach is described in details below.

For the generation of summaries, we used an algorithm that is based on LexRank (Erkan and Radev, 2004), but that solely used the named-entities for the similarity function. In other words, instead of using tf/idf values to rate the importance of each word, we use the named-entities instead.

The first step was to build a sentence graph. Therefore we calculated the cosine similarity of each sentence with each other sentences, i.e., a vector representation of each sentence. However, instead of using each word as dimension for the vector, we only use the named-entities. After the construction of the vectors, we calculate the cosine similarity (cf. equation 1) between each two of these:

$$\text{cosine} = \frac{\sum_{i=1}^{n} A_i B_i}{\sqrt{\sum_{i=1}^{n} A_i^2}\sqrt{\sum_{i=1}^{n} B_i^2}} \qquad (1)$$

where A_i and B_i are the dimensions of the vectors representing the sentences. Afterwards, we create the sentence graph by adding a vertex for each sentence. Then we create edges between those vertices whose corresponding sentences have a similarity score above 0.2.

For calculating the ranking, we used the exact round based formula (cf. equation 2) that is used in LexRank and that originates from PageRank (Page et al., 1999):

$$\text{score}(s_i) = \frac{d}{N} + (1-d) \sum_{s_j \in adj[s_i]} \frac{\text{score}(s_j)}{deg(s_j)} \qquad (2)$$

where N is the total number of vertices in the graph, $adj[s]$ are all adjacent vertices of the vertex s. Additionally, we have the parameter d, a 'damping factor', which is typically set to 0.2 (Page et al., 1999).

Subsequently, we ranked all sentences according to their centrality in the set of related abstracts. We need a last step to generate a summary by removing redundant sentences and we follow the following process:

1. Initialize two sets: (a) an empty set A and a set B that contains all extracted sentences.

2. Order the sentences in set B by decreasing order of their score.

3. Move the top sentence s_i from set B to set A. Then penalize all sentences s_j whose similarity to s_i is greater than a threshold of 0.3 by multiplying their score with the penalty factor of 0.5.

4. Repeat the steps 2 and 3 until enough sentences are in set A.

In a final step, we order the sentences from set A according to their occurrence in the original documents. Thus, we tried to roughly keep the sentence at the position that the author intended.

4 Results and Discussion

In this section, we present the preliminary results we obtained in the fourth edition of the BioASQ challenge. We introduce the details of the BioASQ challenge and then present our results for the two systems with which we participated this year.

4.1 BioASQ challenge

We participated on the Task 4b, which is composed of two phases: A and B. During phase A, the participating teams received a test set of 100 questions along with their question type, i.e., whether yes/no, factoid, list or summary, and had 24 hours to submit their predictions for concepts, documents, passages and RDF triples. When phase A was over, the organizers released the the test set for phase B which contained the same questions previously released for phase A along with gold-standard annotations. During phase B, the participating teams had 24 hours to submit their predictions for exact and ideal answers.

The BioASQ organizers released five bathes of around 100 questions every two weeks. Although our QA systems are capable to output results for most of the tasks covered in BioASQ, we did not submit runs for every sub-task due to problems with the systems, which are still under development.

4.2 Systems

We participated this year with two QA systems, as identified by their run names:

1. HPI1: our previous system that participated in the BioASQ challenge last year (Neves, 2015);

2. HPI2: our new QA system, which is described in this work.

HPI1 is exactly the same system that participated in the BioASQ 2015 and that was one of the winners systems[6]. We made no changes in the

[6] http://www.bioasq.org/participate/third-challenge-winners

system and details on the methods can be found in our previous publication (Neves, 2015). This system was used this year for concept matching and for document and snippet retrieval. The only change made to this system was on the dictionaries which are used in the concept matching task of Phase A. The dictionaries were re-created based on newer versions of the five terminologies specified in the guidelines of the BioASQ challenge: DO, MeSH, Jochem, GO and Uniprot. We downloaded the original files from the respective web sites and compiled dictionaries for each of the terminologies. The dictionaries include various names and synonyms for each concept and was used by the built-in NER functionality of the database to match concepts to the questions.

The document and passage retrieval of the HPI1 system did not make use of our local copy of MEDLINE but it queries PubMed instead. For each question, we generate two queries based on its tokens: (1) by using the "OR" operator and words in the question, except stopwords, and (2) by using the "AND" operator and using all words in the question, except stopwords and words in list of common English words (cf. (Neves, 2015)). We retrieve up to 200 PubMed documents for each of the queries and index these in the IMDB. We rank the sentences for each question based on an approximate similarity between the words in the question and the ones in the document, while a score is automatically calculate between those. Finally, we rank the sentences according to the sum of scores of the matching words and select the top 10 sentences. The list of up to 10 documents is derived from the list to top 10 sentences, i.e., the corresponding documents of these sentences, in the same order.

4.3 Evaluation

Currently, only preliminary results are available for some of the tasks of the BioASQ challenge. We summarize them in Table 1. More details on the results can be found in the BioASQ web site [7].

We present in this section a discussion on the preliminary results that we obtained in the BioASQ challenge, on the limitation of our methods and improvements for future versions of our QA system.

	HPI1	**HPI2**
Concepts	**MAP**	**MAP**
batch1	na	-
batch2	-	-
batch3	na	-
batch4	na	-
batch5	na	-
Documents	**MAP**	**MAP**
batch1	0.0474 (12/15)	0.0028 (15/15)
batch2	-	-
batch3	0.0674 (16/18)	0.0006 (18/18)
batch4	-	-
batch5	0.434 (16/21)	-
Snippets	**MAP**	**MAP**
batch1	0.0481 (1/7)	-
batch2	-	-
batch3	0.0715 (4/14)	-
batch4	-	-
batch5	0.0510 (5/16)	-
Ideal Answ.	**Rouge-2**	**Rouge-2**
batch1	-	0.2231 (1/2)
batch2	-	0.2240 (6/7)
batch3	-	0.2559 (6/7)
batch4	-	0.2280 (4/4)
batch5	-	0.3233 (6/7)

Table 1: Preliminary results in the BioASQ task 4b. Scores for concepts, documents and snippets are in terms of MAP (Mean Average Precision). "na" indicated that results are still not available for this task, while "-" indicated that we did not submit any run for the task. The values inside parameters indicate our current rank and the total number of submissions for the task.

[7] http://participants-area.bioasq.org/results/4b/phaseA/ and http://participants-area.bioasq.org/results/4b/phaseB/

Documents. Curiously, although the strategy used for the document retrieval is exactly the same one used for the snippet retrieval, we obtained much better results for the later, in term of position in the ranking, also in previous editions of the BioASQ challenge. As gold-standard and not available, we can only try to guess the reasons for our performance. When comparing our two systems, HPI2 performed much worse than HPI1, which proves that we still have to need to be improved to deal with large document collections, while HPI1 rely on up to 200 previously retrieved from PubMed.

Snippets. Our system HPI1 performed well again and it a good candidate for obtaining first and second position in the challenge. This proves that the IMDB could effectively match the keywords in the queries to the documents and rank the sentences. However, we see much room for improvement in our approach as named-entities are still not being used in this component, a step which can certainly improve both document and passage retrieval.

Ideal Answers. Our results for ideal answers, i.e., short summaries, provided by system HPI2 also obtained either first or second positions in the all of the batches, when considering results by teams, instead of each individual run.

5 Conclusions and Future Work

In this work we present our results for our two QA systems that participated in task 4b of the BioASQ challenge. The preliminary results show that our approaches are obtained top positions for the snippet retrieval and for the ideal answers. Regarding future work, we envisage much room for improvement for our HPI2 system, the one which is currently under development in our group:

- Both the document and snippet retrieval steps performed much worse than the HPI1 system, which rely on PubMed API. Future work should aim at improving our current ranking algorithms.

- We did not submit runs for factoid and list questions because our system could not return any answer for most of the answers. We did submit one run for yes/no questions but MAP value was of only 25%, while other system are close to 100%. We should perform a comprehensive evaluation of the question processing step, specially the target identification step, and properly integrate further components which can potentially boost our results, such as NER, chunking and semantic role labeling.

Finally, we should perform a comprehensive evaluation on biomedical corpora for the many built-in NLP components of the IMDB, such as NER and POS tagging, as mistakes returned by these can be propagated throughout the system.

Acknowledgments

We thank David Heller, Thomas Hille and Fabian Eckert for the interesting discussions during the project and the HPI Future Soc Lab Team for providing us the access to the IMDB. Finally, we also would like to thank technical support from the students of our current Bachelor Project: Maximilian Goetz, Marcel Jankrift, Julian Niedermeier, Toni Stachewicz and Soeren Tietboehl.

References

Sofia J. Athenikos and Hyoil Han. 2010. Biomedical question answering: A survey. *Computer Methods and Programs in Biomedicine*, 99(1):1 – 24.

MichaelA Bauer and Daniel Berleant. 2012. Usability survey of biomedical question answering systems. *Human Genomics*, 6(1):1–4.

Günes Erkan and Dragomir R Radev. 2004. Lexrank: Graph-based lexical centrality as salience in text summarization. *Journal of Artificial Intelligence Research*, pages 457–479.

Zhiheng Huang, Marcus Thint, and Zengchang Qin. 2008. Question classification using head words and their hypernyms. In *Proceedings of the Conference on Empirical Methods in Natural Language Processing*, EMNLP '08, pages 927–936, Stroudsburg, PA, USA. Association for Computational Linguistics.

Daniel Jurafsky and James H. Martin. 2013. *Speech and Language Processing*. Prentice Hall International, 2 revised edition.

Christopher D. Manning, Prabhakar Raghavan, and Hinrich Schütze. 2008. *Introduction to Information Retrieval*. Cambridge University Press, Cambridge, UK.

Mariana Neves and Ulf Leser. 2015. Question answering for biology. *Methods*, 74:36 – 46.

Mariana Neves. 2015. HPI question answering system in the bioasq 2015 challenge. In *Working Notes of CLEF 2015 - Conference and Labs of the Evaluation forum, Toulouse, France, September 8-11, 2015.*

Lawrence Page, Sergey Brin, Rajeev Motwani, and Terry Winograd. 1999. The pagerank citation ranking: bringing order to the web.

Hasso Plattner. 2013. *A Course in In-Memory Data Management: The Inner Mechanics of In-Memory Databases*. Springer, 1st edition.

George Tsatsaronis, Georgios Balikas, Prodromos Malakasiotis, Ioannis Partalas, Matthias Zschunke, Michael R Alvers, Dirk Weissenborn, Anastasia Krithara, Sergios Petridis, Dimitris Polychronopoulos, et al. 2015. An overview of the bioasq large-scale biomedical semantic indexing and question answering competition. *BMC bioinformatics*, 16(1):138.

KSAnswer: Question-answering System of Kangwon National University and Sogang University in the 2016 BioASQ Challenge

Hyeon-gu Lee[1], **Minkyoung Kim**[1], **Harksoo Kim**[1*], **Juae Kim**[2]
Sunjae Kwon[2], **Jungyun Seo**[2*], **Jungkyu Choi**[3], **Yi-reun Kim**[3]
Kangwon National University, Chuncheon, Korea[1]
Sogang University, Seoul, Korea[2]
Intelligence Lab, LG Electronics, Korea[3]
{nlphglee, kmink0817, nlpdrkim}@kangwon.ac.kr[1]
{juaeKim, sj91kwon, seojy}@sogang.ac.kr[2]
{stanley.choi, yireun.kim }@lge.com[3]

Abstract

This paper describes a question–answering system that returns relevant documents and snippets (with particular emphasis on snippets) from a large medical document collection. The system is implemented as part of our participation to Phase A of Task 4b in the 2016 BioASQ Challenge. The proposed system retrieves candidate answer sentences using a cluster–based language model. Then, it re–ranks the retrieved top-n sentences using five independent similarity models based on shallow semantic analysis. The experimental results show that the proposed system is the first to find snippets in batches 2 (MAP 0.0604), 3 (MAP 0.0728), 4 (MAP 0.1182), and 5 (MAP 0.1187).

1 Introduction

BioASQ 2016 is the fourth annual BioASQ challenge as an established international competition for large–scale biomedical semantic indexing and question–answering, since 2013 (Tsatsaronis et al., 2015). The challenge consists of two tasks: Task 4a on large–scale online biomedical semantic indexing and Task 4b on biomedical semantic question–answering. Task 4b is further divided into two phases: Phase A and Phase B. In Phase A, participating systems are required to return a maximum of 10 relevant concepts, documents, snippets, and triples during five batches. Participation in Phase A can be partial, which means that it is acceptable to participate in only some of the batches and to return only relevant documents without snippets, triples, and concepts. This paper describes a questionanswering system of Kangwon National University and Sogang University submitted for Phase A of Task 4b in BioASQ 2016. The proposed system is focused on returning relevant documents and snippets (with particular emphasis on snippets).

2 Question-answering system based on sentence retrieval and re–ranking techniques

KSAnswer consists of two submodules: A retrieval model for finding candidate answer sentences from a large medical collection and a re–ranking model for determining the final answer among the retrieved candidate answer sentences.

2.1 Sentence retrieval model

Prior to indexing documents, KSAnswer first splits documents into a sequence of sentences using LingPipe (Baldwin et al., 2003). Then, it performs morphological analysis of the sentences and extracts content words (i.e., proper noun, common noun, verb, number, and so on) from the sentences. This is followed by stemming of content words except proper nouns using Porter Stemmer (Porter, 1980). Finally, KSAnswer uses the stemmed content words and the proper nouns as indexing terms.

For cluster–based sentence retrieval, KSAnswer generates two types of indexing units from the document collection comprising full data sets of PubMed journals: a sentence trigram unit and a document unit. The sentence trigram unit consists of an indexing target sentence and its context sentences (the previous and the next sentences) to address the lexical disagreement between a query and an indexed sentence. If a document consists of three sentences, KSAnswer generates three sentence trigrams (NULL–1st sentence–2nd sentence, 1st sentence–2nd sentence–3rd sentence,

*Corresponding author.

Proceedings of the 54th Annual Meeting of the Association for Computational Linguistics, pages 45–49,
Berlin, Germany, August 7-12, 2016. ©2016 Association for Computational Linguistics

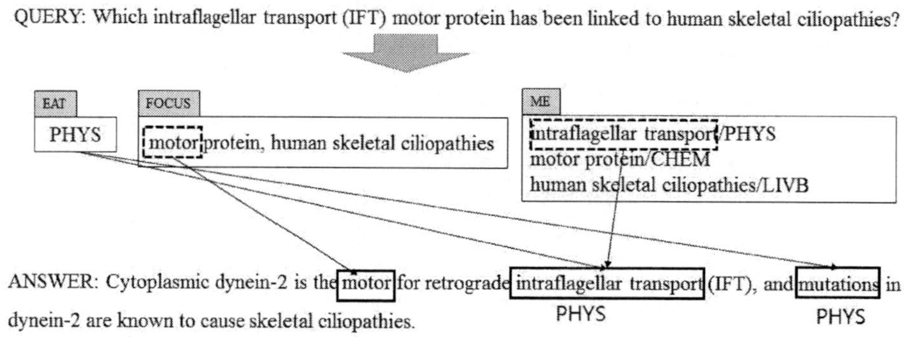

Figure 1: Relationship between a query and an answer sentence

2nd sentence–3rd sentence–NULL). The document unit consists of a title sentence and abstract sentences. The document unit assists in addressing the lexical disagreement between a query and a sentence trigram. Then, KSAnswer performs indexing of each unit and constructs two indexing databases using Lucene 4.0.0 (Białecki et al., 2012).

To rank candidate answer sentences, KSAnswer uses a cluster–based language model (Liu et al., 2004; Merkel et al., 2007), as shown in Eq. (1):

$$Sim_{IR}(Q,S) = \propto Sim_{tri}(Q,T) + (1-\propto) Sim_{doc}(Q,D) \quad (1)$$

where $Sim_{tri}(Q,T)$ is the similarity of the language model between the query Q and the sentence trigram T in the document D. Then, $Sim_{doc}(Q,D)$ is the similarity of the language model between the query Q and the document D. The weighting parameter $\propto$ has a value between 0 and 1. Finally, $Sim_{IR}(Q,S)$ returns similarities between the query Q and the indexing target sentence S, which is located in the middle of the sentence trigram T.

2.2 Sentence re–ranking model

Prior to re–ranking of candidate answer sentences, KSAnswer selects top–n retrieved sentences and normalizes their similarities, as shown in Eq. (2):

$$Sim'_{IR}(Q,S) = \frac{Sim_{IR}(Q,S) - m}{\sigma} \quad (2)$$

where m and σ are the average and standard deviation of similarity scores of top–n retrieved sentences, respectively.

KSAnswer re–ranks the top–n retrieved sentences using five independent similarity models, namely, $Sim_{SNT}(Q,S)$, $Sim_{EMB}(Q,S)$, $Sim_{EAT}(Q,S)$, $Sim_{FOCUS}(Q,S)$, and $Sim_{ME}(Q,S)$. $Sim_{SNT}(Q,S)$ is a similarity model between the query Q and the sentence S, which is located in the middle of the retrieved sentence trigram. $Sim_{EMB}(Q,S)$ is a similarity model between the sentence embedding of Q and the sentence embedding of S. The sentence embeddings are constructed by the sum of position–encoded word vectors in Word2Vec (so–called position encoding) (Sukhbaatar et al., 2015). $Sim_{EAT}(Q,S)$ is a similarity model between the expected answer type (EAT; a category name of expected answer) of Q and medical entity types (category names of medical entities) in S. $Sim_{FOCUS}(Q,S)$ is a similarity model between focus words (FOCUS; a clue word sequence to identify correct answers) in Q and content words in S. $Sim_{ME}(Q,S)$ is a similarity model between medical entities (MEs) in Q and medical entities in S. For example, in the sentence "Which drugs are utilized to treat eosinophilic esophagitis?", EAT, FOCUS, and ME are *[Chemicals & Drugs]*, *[eosinophilic esophagitis]*, and *[drugs, eosinophilic esophagitis]*, respectively. To obtain EAT, FOCUS, and ME, KSAnswer uses a sentence analyzer based on pattern matching and machine learning (Kim et al., 2004). The sentence analyzer extracts word chunks (generally noun phrases) from a query using lexico–semantic patterns. Then, it determines EAT and FOCUS by searching the syntactic chunks in MetaMap (Aronson et al., 2006). To obtain MEs, the sentence analyzer uses a special version of named entity recognizer based on Conditional Random Fields (CRFs), which is trained for medical documents (Abacha

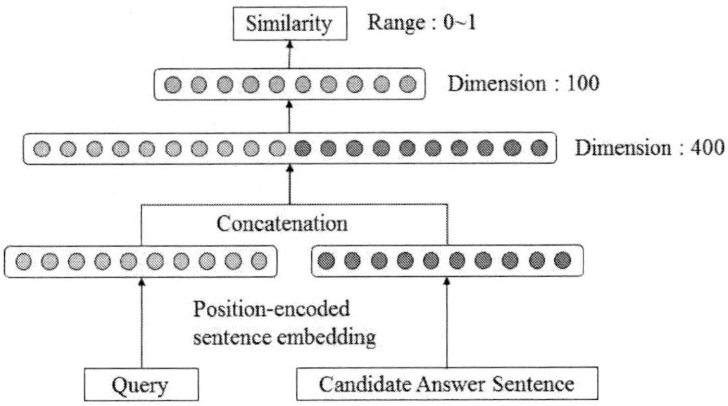

Figure 2: Vector-based similarity model based on a neural network

et al., 2012). The named entity recognizer extracts medical entities from a sentence and annotates them with predefined semantic categories. EAT and MEs use the same semantic categories as follows: ACTI (Activities & Behaviors), ANAT (anatomy), CHEM (chemicals & drugs), CONC (concepts & ideas), DEVI (devices), DISO (disorders), GENE (genes & molecular sequences), GEOG (geographic areas), LIVB (living beings), OBJC (objects), OCCU (occupations), ORGA (organizations), PHEN (phenomena), PHYS (physiology), and PROC (procedures). Figure 1 shows a relationship between Q and S at the view of EAT, FOCUS, and ME.

Eq. (3) shows the similarity scores between a query and each top–*n* retrieved sentences for re–ranking.

$$\begin{aligned} ReSim(Q,S) &= \alpha Sim_{IR}^{'} \\ &+(1-\alpha)\{\beta Sim_{snt}(Q,S) \\ &+(1-\beta)\sum_{i=1}^{4}\gamma_i Sim_{sem}^{i}(Q,S)\}, \\ where\ 0 \le \alpha \le 1, &0 \le \beta \le 1, \sum_{i=1}^{4}\gamma_i = 1 \end{aligned} \quad (3)$$

where $Sim_{sem}^{i}(Q,S)$ is the ith similarity model among $Sim_{EMB}(Q,S)$, $Sim_{EAT}(Q,S)$, $Sim_{FOCUS}(Q,S)$, and $Sim_{ME}(Q,S)$. Then, $\propto$, β, and γ are the weighting parameters set by experiments. The word–based similarity models (i.e., models for calculating similarities between words in Q and S), such as $Sim_{SNT}(Q,S)$, $Sim_{FOCUS}(Q,S)$, and $Sim_{ME}(Q,S)$, are calculated using the well-known Okapi BM25 (Robertson et al., 1999). Then, the category–based similarity model (i.e., a model for calculating similarities between category names in Q and S), $Sim_{EAT}(Q,S)$, is calculated using OR similarity of the Paice model (Paice, 1984), as shown in Eq. (4).

$$\begin{aligned} Sim_{EAT}(Q,S) &= \frac{\sum_{i=1}^{n}(r^{i-1}w_i)}{\sum_{i=1}^{n}r^{i-1}}, \\ where\ 0 \le r \le 1\ &and\ w_i^{'}s\ are \\ considered\ in\ &descending\ order \end{aligned} \quad (4)$$

In Eq. (4), w_i is a TF·IDF value of the ith word in ME's of S that have the same semantic category with EAT of Q. Finally, the vector–based similarity model, $Sim_{EMB}(Q,S)$, is calculated using a feed–forward neural network with one hidden layer (Svozil et al., 1997), as shown in Figure 2.

The feed–forward neural network uses the sentence embedding vectors of Q and S as input values and uses a degree of relevance (from 0 to 1) between the two sentence embedding vectors as an output value. It is trained using gold standard answers as relevant snippets and by using top–*n* retrieved sentences except gold standard answers as irrelevant snippets.

3 Experiments

3.1 Experimental setting

We indexed the full data set of PubMed journals using Lucene 4.0.0. The number of document units was 12,208,342 and the number of sentence trigram units was 99,911,516. The language model parameters (μ values) for the document and sentence trigram units were set to 500 and 100, respectively. The weighting parameter $\propto$ in Eq. (1)

was 0.8. Then, the weighting parameters $\propto$, β, and γ_i in Eq. (3) were 0.5, 0.9, and 0.3, respectively.

3.2 Experimental results

In Phase A of Task 4b, our best submission was the first to find snippets in batches 2, 3, 4, and 5. In batch 1, we indexed the limit set of PubMed and achieved the second place in finding snippets. Table 1 shows the best performances of KSAnswer.

Table 1: Evaluation results of submitted runs

Batch	Document	
	Precision	Recall
	F1	MAP
1	0.0840(0.0840)	0.2258(0.1664)
	0.1065(0.1116)	0.0486(0.1223)
2	0.1675(0.1675)	0.4056(0.2758)
	0.2122(0.2084)	0.0949(0.1905)
3	0.1380(0.1380)	0.3946(0.2686)
	0.1786(0.1823)	0.0992(0.2095)
4	0.1720(0.1720)	0.5333(0.3470)
	0.2247(0.2300)	0.1257(0.2871)
5	0.1103(0.1103)	0.3752(0.2560)
	0.1546(0.1542)	0.0752(0.1742)
Batch	Snippet	
	Precision	Recall
	F1	MAP
1	0.0482(0.0418)	0.0952(0.1071)
	0.0534(0.0602)	0.0266(0.0738)
2	0.1021(0.0967)	0.1615(0.1930)
	0.1104(0.1288)	0.0604(0.1381)
3	0.0873(0.0823)	0.1208(0.1460)
	0.0886(0.1053)	0.0728(0.1440)
4	0.1504(0.1377)	0.2023(0.2653)
	0.1554(0.1813)	0.1182(0.2549)
5	0.0771(0.0773)	0.1272(0.1434)
	0.0798(0.1004)	0.0582(0.1187)

The parenthesized values are informal performances that are calculated using gold standard answers for each batch. In an additional experiment, we found that the degree of the sub-model importance in the reranking model is as follows: $Sim_{SNT}(Q,S)$ $>>$ $Sim_{EAT}(Q,S)$ $>$ $Sim_{FOCUS}(Q,S)$ $\approx$ $Sim_{ME}(Q,S) \approx Sim_{EMB}(Q,S)$

4 Conclusion

We proposed a question-answering system for finding candidate answer snippets from a large medical document collection. The proposed system retrieves candidate answer sentences using cluster–based language model. Then, it re–ranks top–*n* retrieved sentences using various similarity models based on shallow semantic analysis of sentences. In Phase A of task 4b, the proposed system showed excellent performance by being the first to find snippets in batches 2,3,4 and 5.

Acknowledgments

This research was supported by LG Electronics. It was also supported by Basic Science Research Program through the National Research Foundation of Korea (NRF) funded by the Ministry of Education, Science and Technology (2013R1A1A4A01005074).

References

Alan R. Aronson. 2006. Metamap: Mapping text to the umls metathesaurus. *Bethesda*, MD: NLM, NIH, DHHS, 1–26.

Andreas Merkel, and Dietrich Klakow. 2007. Comparing improved language models for sentence retrieval in question answering. *LOT Occasional Series 7*, 35–50.

Andrzej Białecki, Robert Muir, and Grant Ingersoll. 2012. Apache lucene 4. *SIGIR 2012 workshop on open source information retrieval.*

Ben Abacha, Asma, and Pierre Zweigenbaum. 2012. Medical question answering: translating medical questions into sparql queries. *Proceedings of the 2nd ACM SIGHIT International Health Informatics Symposium.* ACM.

Breck Baldwin, and Bob Carpenter. 2003. LingPipe. *Available from World Wide Web: http://alias-i.com/lingpipe.*

Chris D. Paice. 1984. Soft evaluation of Boolean search queries in information retrieval systems. *Information Technology: Research and Development,* 3(1):33-41.

Daniel Svozil, Vladimir Kvasnicka, and Jiri Pospichal. 1997. Introduction to multi–layer feed-forward neural networks. *Chemometrics and intelligent laboratory systems,* 39(1):43–62.

George Tsatsaronis, Georgios Balikas, Prodromos Malakasiotis, Ioannis Partalas, Matthias Zschunke, Michael R. Alvers, Dirk Weissenborn, Anastasia Krithara, Sergios Petridis, Dimitris Polychronopoulos, Yannis Almirantis, John Pavlopoulos, Nicolas Baskiotis, Patrick Gallinari, Thierry Artires, Axel–Cyrille N. Ngomo, Norman Heino, Eric Gaussier,

Liliana Barrio-Alvers, Michael Schroeder, Ion Androutsopoulos and Georgios Paliouras. 2015. An overview of the BIOASQ large–scale biomedical semantic indexing and question answering competition. *BMC Bioinformatics.*

Harksoo Kim and Jungyun Seo. 2004. A high performance question-answering system based on a two–pass answer indexing and lexico-syntactic pattern matching. *IEICE Information and Systems,* Vol.E87–D (12):2855–2862.

Martin F. Porter. 1980. An algorithm for suffix stripping. *Program,* 14(3):130–137.

Sainbayar Sukhbaatar, Jason Weston, and Rob Fergus. 2015. End–to–end memory networks. *Advances in Neural Information Processing Systems.*

Stephen E. Robertson, Steve Walker, and M. Beaulieu. 1999. Okapi at TREC–7: automatic ad hoc, filtering, VLC and interactive track. *Nist Special Publication SP,* 253–264.

Xiaoyong Liu, and W. Bruce Croft. 2004. Cluster–based retrieval using language models. *Proceedings of the 27th annual international ACM SIGIR conference on Research and development in information retrieval.* ACM.

Large-Scale Semantic Indexing and Question Answering in Biomedicine

E. Papagiannopoulou*, Y. Papanikolaou*, D. Dimitriadis*, S. Lagopoulos*, G. Tsoumakas*, M. Laliotis**, N. Markantonatos** and I. Vlahavas*

*School of Informatics, Aristotle University of Thessaloniki, Thessaloniki 54124, Greece
**Atypon, 5201 Great America Parkway Suite 510, Santa Clara, CA 95054, USA

Abstract

In this paper we present the methods and approaches employed in terms of our participation in the 2016 version of the BioASQ challenge. For the semantic indexing task, we extended our successful ensemble approach of last year with additional models. The official results obtained so-far demonstrate a continuing consistent advantage of our approaches against the National Library of Medicine (NLM) baselines. For the question answering task, we extended our approach on factoid questions, while we also developed approaches for the document, concept and snippet retrieval sub-tasks.

1 Introduction

The BioASQ project (Balikas et al., 2014) aims to provide a challenge framework for researchers dealing with classification (semantic indexing) and natural language processing (question answering) tasks in the field of bio-medicine. The challenge, similar to the previous three years, is divided into two tasks: automated semantic indexing (4A) and question answering (4B).

In Task 4A participants are given a set of new, unannotated articles and are required to automatically predict the relevant MeSH terms for each one of them in a given time. For each article only the abstract along with some meta-information is provided (journal, year and title). This task is particularly difficult, as the MeSH taxonomy is comprised of a large number of labels (~ 27000), with the label set following a distribution similar to power-law. Furthermore the terms are subject to a significant concept drift along time.

Task 4B is divided into 2 phases, called A and B. In phase A participants are given a set of questions and must return the 10 most relevant documents, snippets, concepts (from designated ontologies) and RDF triples. In phase B participants are given the gold standard documents and snippets and must provide exact and ideal answers.

This paper discusses the approaches we developed for this year's BioASQ challenge. In particular, Section 2 discusses our semantic indexing algorithms, Section 3 our document retrieval system, Section 4 our concept retrieval method, Section 5 our snippet retrieval approach and Section 6 discusses our question answering approach. Final considerations and conclusions are drawn in Section 7.

2 Task 4A: Semantic Indexing

In this section we present the methods that we used for the semantic indexing task. We first provide the pre-processing pipeline and subsequently the methods employed.

2.1 Pre-processing

In this year's participation, we used the 1,050,000 most recent documents from the BioASQ 2016 corpus using as a training set the first 1 million articles and the last 50 thousand as a validation set. The motivation behind using the latest articles of the corpus, stems from the hypothesis that more recent chronologically articles will tend to follow more similar labels distributions to new articles that have to be predicted, compared to older ones. Pre-processing of the articles was carried out similar to previous years; the abstract and the title were concatenated, uni-grams and bi-grams were used as features, removing stop-words and features with less than five occurrences in the corpus. We used the *tf-idf* representation for the features. Also, zoning of the features belonging to the title and those equal to a MeSH label was performed

Proceedings of the 54th Annual Meeting of the Association for Computational Linguistics, pages 50–54,
Berlin, Germany, August 7-12, 2016.

by increasing the *tf-idf* value of features that belonged to the title by $log2$ and those being equal to a label by $log1.25$. The above features were used in order to train several multi-label learning models, described in the following section.

2.2 Methods

Our participation to this year's contest included several multi-label classifiers (MLC) that were combined in various ensembles. As in the previous year, we used the Meta-Labeler (Tang et al., 2009), a set of Binary Relevance (BR) models with Linear SVMs (both tuned and with default parameters) and a Labeled LDA variant, Prior LDA (Rubin et al., 2012). For the tuned SVM models, we used different values for the C parameter and handled class imbalance by penalizing more heavily false negative errors than false positive ones by adjusting properly the weight parameter (Lewis et al., 2004). This year, we additionally employed Fast XML (Prabhu and Varma, 2014) and HOMER-BR (Tsoumakas et al., 2008).

All the above models were combined in an ensemble, using the MULE framework (Papanikolaou et al., 2014). MULE is a statistical significance multi-label ensemble that performs classifier selection. The key idea is to combine a set of multi-label classifiers aiming to optimize a selected measure (for the purpose of this challenge, we are mainly interested in the micro-F measure) and validate this combination through a statistical significance test; McNemar's test. This way, each label of the multi-label problem is predicted with a specific component model, the one that (a) contributes to the greatest improvement to the evaluation metric of interest and (b) is validated from the statistical test to indeed produce the aforementioned improvement. If (b) does not hold, in other words if the component model's improvement is not statistically significant, we predict that label with the globally optimal model.

2.3 Results

Since at the moment of writing this paper there are not sufficient official results yet(only the a small part of documents of the first batch are annotated), in Table 1 we present the performance of the multi-label classifiers used in our ensembles, in terms of the Micro-F and Macro-F measures, for the training set (one million documents) and the validation set (fifty thousand documents) used throughout the challenge.

Table 1: Performance of the multi-label classifiers used throughout the BioASQ challenge semantic indexing task 4a, in terms of Micro-F and Macro-F. Training set size was 1,000,000 documents and test set size 50,000 respectively.

MLC	Micro-F	Macro-F
Meta-Labeler	**0.61936**	**0.57477**
Vanilla SVMs	0.58422	0.50080
Tuned SVMs	0.61365	0.54444
Labeled LDA	0.47399	0.39084
Fast XML	0.38053	0.28899
HOMER-BR (k=3)	0.59698	0.54972

3 Task 4B Phase A: Documents

Here we describe our document retrieval system. The system was written in Java. A variety of libraries have been used. The StAX Parser[1] for the input of XML files, the Stanford Parser[2] for natural language parsing and the GSON library[3] for output of JSON files. We build our system on open source Indri search engine from the Lemur Project[4].

3.1 Pre-processing of citations

We processed the full database of MEDLINE and extracted the citations that contained Title, Abstract and MeSH annotations. There are 14,938,869 documents.

3.2 Search Engine

We used Indri as our search engine. We normalized the text of all the processed citations and we inserted them to our search engine. No stemming or stop-words filtering has been done in order to avoid any distortion of bio-medical and other important terminology.

3.3 Question Parsing and Query

Our system processes and analyzes the input question before producing the final query. It removes any unwanted punctuation, it analyzes the question with the Stanford Parser and produces a bag of words. Finally, we form our query by combining the bag of words with the query language grammar of Indri.

[1] https://docs.oracle.com/javase/tutorial/jaxp/stax/api.html

[2] http://nlp.stanford.edu/software/lex-parser.shtml

[3] https://github.com/google/gson

[4] http://www.lemurproject.org/

3.4 Testing

We tested our system by using both the questions and the gold standard articles of the previous BioASQ challenges and the current challenge. We experimented with Indri's great variety of search terms and tried retrieving top-10, top-20 and top-50 documents. The table below provides the results of our experiments retrieving top-10 documents.

Table 2: Test results retrieving top-10 documents

Task	# questions	Precision	MAP
1b, 2b, 3b	940	0.279	0.141
4b TestSet 1	100	0.156	0.233
4b TestSet 2	100	0.230	0.198
4b TestSet 3	100	0.195	0.250
4b TestSet 4	100	0.235	0.321
4b TestSet 5	97	0.105	0.158

4 Task 4B Phase A: Concepts

We are working at the phase A task of returning a list of at most 10 relevant concepts from the designated terminologies and ontologies. The list is ordered by decreasing confidence. In our approach, we use MetaMap[5] and LingPipe[6] to detect the biomedical concepts and local ontology files (Disease ontology, Gene ontology, Jochem, Uniprot and MeSH) to retrieve the appropriate information. More particularly, we use RDF4J[7], a powerful Java framework for processing and handling RDF data of Disease ontology, Gene ontology, Jochem, and MeSH. This includes creating, parsing, storing, inferencing and querying over such data. Additionally, we use RDF4J's Lucene Sail that enables us to add full text search of RDF literals to find fast subject resources. As far as the Uniprot data are concerned which are not in obo format, we exploit them in XML format (not plain text that is recommended by the contest). Of course, Lucene indexing is necessary again. We present our methodology step by step:

1. The first step of our methodology is to remove stopwords from the given question. We use 2 stopwords lists: a basic list with 634 words and the Pubmed stopword list[8]. Then, we detect keywords using MetaMap and LingPipe. We give a boosting score to those concepts that come from MetaMap/LingPipe and a smaller score in any other word that appears in the question and MetaMap/LingPipe does not recognize it as biomedical concept.

2. Then, we expand the list with the candidate concepts exploiting the MeSH ontology (up to 15 candidate concepts, totally, enriching the list with ExactSynonyms). We retain two lists with candidate concepts: a list with all possible biomedical concepts for search in Disease ontology, Gene ontology, Jochem, and MeSH and a list that contains only proteins or genes for search in Uniprot XML data.

3. We search for each candidate term separately combining search in RDF4J's Lucene Sail index for fast detection of relevant terms and search in RDF4J RDF Repositories via SPARQL queries to filter the results which are returned as relevant terms by RDF4J's Lucene Sail index. More specifically, for the 4 ontologies we examine if the candidate term appears in properties: label, ExactSynonym, RelSynonym, Synonym, NarrowSynonym, BroadSynonym in order to add to Lucene score an additional boosting score and return the corresponding URI. If the candidate term does not appear in the above properties, then we just keep the Lucene score. Additionally, we exploit the properties (Positively/Negatively) Regulates in order to return the corresponding URI, too. Similarly, we conduct search in Uniprot data but instead of SPARQL queries, we use XPath, focusing in the following XML elements: fullName, shortName, alternativeName and innName.

4. Finally, we take the top 10 concepts with the biggest scores.

Here, we present experimental results on 2 different sets of questions (the sets belong to the training set of BioASQ contest).

Table 3: Results of our approach

# questions	Precision	Recall	F1	MAP
238	0.167	0.511	0.223	0.120
286	0.209	0.513	0.267	0.167

[5]https://metamap.nlm.nih.gov/

[6]http://alias-i.com/lingpipe/

[7]http://rdf4j.org/

[8]http://www.ncbi.nlm.nih.gov/books/NBK3827/table/pubmedhelp.T.stopwords/

5 Task 4B Phase A: Snippets

In order to extract relevant snippets to a query, we exploit our knowledge given by the ontologies we referred to in Section 4 (Disease ontology, Gene ontology, Jochem, Uniprot and MeSH). Briefly:

1. Detect keywords using MetaMap and LingPipe

2. Search for synonyms for each keyword in order to make query expansion. Consider we have K keywords and for each one we find a few synonyms, e.g. for i-th keyword, $i = 1, ..., K$, we detect N synonyms. Each synonym is denoted by $synj_{key_i}$, that is the j-th synonym (syn_j), $j = 1, ..., N$ of i-th keyword (key_i).
   Format of query after the expansion step: Suppose K=2, key_1 has N synonyms and key_2 has M synonyms, so the query is going to be the following:
   ((key_1 OR $syn1_{key_1}$ OR $syn2_{key_1}$ OR ... OR $synN_{key_1}$)
   AND
   (key_2 OR $syn1_{key_2}$ OR $syn2_{key_2}$ OR ... OR $synM_{key_2}$))
   The total number of the candidate concepts (i.e. keywords with their corresponding synonyms) should contain up to 15 concepts.

3. Retrieve top 100 relevant documents (use of Lucene index). More particularly, we are interested in their title, abstract and pmid.

4. Split titles/abstracts returned in step 3 into sentences.

5. Calculate sematic similarity between each one of the sentences and the (expanded) query using the semantic similarity measure described in (Han et al., 2013). (At this point, we experiment using clustering algorithms in order to select the sentences that are located in the same cluster with the query, regarding them as the most relevant snippets.)

6. Return the top 10 sentences that are more similar to our query according to the similarity measure.

6 Task 4B, Phase B: Exact Answers

We developed a system that extracts answers from factoid questions under a scoring mechanism. In our approach, we applied numerous measurements that rank the candidate answers based on their relations with the questions. Some of them were applied in our previous system, but we realized that were not enough to estimate the correct answer. Thus, we extended the previous scoring mechanism in order to include the measures describing below.

- *distance:* The words, being near to the LAT of the question into the snippets, it is more possible to be a candidate answer.

- *wordnet synonyms:* We strongly believe that words with many synonyms in wordnet are more likely to be used in common language rather than in biomedicine. Thus, they take a punishment according to the number of synonyms that they have.

Furthermore, in the previous work, the system selected some of the words of an article as candidate answers. It selected the words that were produced by MetaMap parsing. Although, the results of the previous system were promising in the BioASQ training set, in the BioASQ challenge were quite low. The system's failure was caused by the lack of candidate answers. That's why we decided to expand the set of candidate answers considering all words including in the related snippets of a question.

Finally, the specificity measure in our previous work changed because of the execution time. We had implemented that measure to count the number of instances of a candidate answer in all PubMed documents. Thus, we decided to seek the documents including the candidate answers with a document retrieval system. For each retrieving document, the candidate answer take a punishment.

Table 4: Results of factoid system

LACC	SACC	MRR
0.54	0.237	0.305

7 Conclusions

In this paper we presented the participation of our team in the BioASQ challenge 2016. Building on the successful approaches in the past three challenges, we further extended our line of work to improve the performance of our systems. Additionally, our methodology for relevant concepts retrieval gives quite good results based on our evaluation in a variety of bio-medical questions that are provided by BioASQ's training set. Moreover, the semantic information from ontologies could be exploited for other tasks.

References

[Balikas et al.2014] Georgios Balikas, Ioannis Partalas, Axel-Cyrille Ngonga Ngomo, Anastasia Krithara, and Georgios Paliouras. 2014. Results of the bioasq track of the question answering lab at CLEF 2014. In *Working Notes for CLEF 2014 Conference, Sheffield, UK, September 15-18, 2014.*, pages 1181–1193, july.

[Han et al.2013] Lushan Han, Abhay L. Kashyap, Tim Finin, James Mayfield, and Johnathan Weese. 2013. UMBC_EBIQUITY–CORE: Semantic Textual Similarity Systems. In *In Proceedings of the 2nd Joint Conf. on Lexical and Computational Semantics*. Association for Computational Linguistics.

[Lewis et al.2004] David D. Lewis, Yiming Yang, Tony G. Rose, and Fan Li. 2004. Rcv1: A new benchmark collection for text categorization research. *J. Mach. Learn. Res.*, 5:361–397.

[Papanikolaou et al.2014] Yannis Papanikolaou, Dimitrios Dimitriadis, Grigorios Tsoumakas, Manos Laliotis, Nikos Markantonatos, and Ioannis P. Vlahavas. 2014. Ensemble approaches for large-scale multi-label classification and question answering in biomedicine. In *Working Notes for CLEF 2014 Conference, Sheffield, UK, September 15-18, 2014.*, pages 1348–1360.

[Prabhu and Varma2014] Yashoteja Prabhu and Manik Varma. 2014. Fastxml: a fast, accurate and stable tree-classifier for extreme multi-label learning. In *Proceedings of the 20th ACM SIGKDD international conference on Knowledge discovery and data mining*, pages 263–272. ACM.

[Rubin et al.2012] Timothy N. Rubin, America Chambers, Padhraic Smyth, and Mark Steyvers. 2012. Statistical topic models for multi-label document classification. *Mach. Learn.*, 88(1-2):157–208, July.

[Tang et al.2009] Lei Tang, Suju Rajan, and Vijay K. Narayanan. 2009. Large scale multi-label classification via metalabeler. In *WWW '09: Proceedings of the 18th international conference on World wide web*, pages 211–220, New York, NY, USA. ACM.

[Tsoumakas et al.2008] G. Tsoumakas, I. Katakis, and I. Vlahavas. 2008. Effective and efficient multilabel classification in domains with large number of labels. In *Proc. ECML/PKDD 2008 Workshop on Mining Multidimensional Data (MMD'08)*, pages 30–44.

Author Index

Association for Computational Linguistics
209 N. Eighth Street
Stroudsburg, Pennsylvania 18360

CURRAN ASSOCIATES INC.
proceedings.com

ISBN 978-1-5108-2764-6

9 781510 827646